United States Department of Agriculture

Changing Climate, Changing Forests: The Impacts of Climate Change on Forests of the Northeastern United States and Eastern Canada

Lindsey Rustad, John Campbell, Jeffrey S. Dukes, Thomas Huntington, Kathy Fallon Lambert, Jacqueline Mohan, and Nicholas Rodenhouse

Abstract

Decades of study on climatic change and its direct and indirect effects on forest ecosystems provide important insights for forest science, management, and policy. A synthesis of recent research from the northeastern United States and eastern Canada shows that the climate of the region has become warmer and wetter over the past 100 years and that there are more extreme precipitation events. Greater change is projected in the future. The amount of projected future change depends on the emissions scenarios used. Tree species composition of northeast forests has shifted slowly in response to climate for thousands of years. However, current human-accelerated climate change is much more rapid and it is unclear how forests will respond to large changes in suitable habitat. Projections indicate significant declines in suitable habitat for spruce-fir forests and expansion of suitable habitat for oak-dominated forests. Productivity gains that might result from extended growing seasons and carbon dioxide and nitrogen fertilization may be offset by productivity losses associated with the disruption of species assemblages and concurrent stresses associated with potential increases in atmospheric deposition of pollutants, forest fragmentation, and nuisance species. Investigations of links to water and nutrient cycling suggest that changes in evapotranspiration, soil respiration, and mineralization rates could result in significant alterations of key ecosystem processes. Climate change affects the distribution and abundance of many wildlife species in the region through changes in habitat, food availability, thermal tolerances, species interactions such as competition, and susceptibility to parasites and disease. Birds are the most studied northeastern taxa. Twenty-seven of the 38 bird species for which we have adequate long-term records have expanded their ranges predominantly in a northward direction. There is some evidence to suggest that novel species, including pests and pathogens, may be more adept at adjusting to changing climatic conditions, enhancing their competitive ability relative to native species. With the accumulating evidence of climate change and its potential effects, forest stewardship efforts would benefit from integrating climate mitigation and adaptation options in conservation and management plans.

Cover Photo

Sandy Stream Pond at the foot of Mt. Katahdin, Baxter State Park, Maine. Photo by PaulCyrPhotography.com.

Manuscript received for publication 22 September 2011

Published by:

U.S. FOREST SERVICE
11 CAMPUS BLVD SUITE 200
NEWTOWN SQUARE PA 19073

July 2012

For additional copies:

U.S. Forest Service
Publications Distribution
359 Main Road
Delaware, OH 43015-8640
Fax: (740)368-0152
Email: nrspubs@fs.fed.us

Visit our homepage at: **http://www.nrs.fs.fed.us/**

Changing Climate, Changing Forests: The Impacts of Climate Change on Forests of the Northeastern United States and Eastern Canada

LINDSEY RUSTAD is a research ecologist with the U.S. Forest Service, Northern Research Station, Durham, NH.

JOHN CAMPBELL is a research ecologist with the U.S. Forest Service, Northern Research Station in Durham, NH.

JEFFREY S. DUKES is an associate professor of forestry and natural resources at Purdue University, West Lafayette, IN.

THOMAS HUNTINGTON is a research hydrologist with the U.S. Geological Survey in Augusta, ME.

KATHY FALLON LAMBERT is a science and policy program director with Harvard Forest, in Petersham, MA.

JACQUELINE MOHAN is an assistant professor at the Odum School of Ecology, University of Georgia, Athens, GA.

NICHOLAS RODENHOUSE is a professor of biological sciences at Wellesley College, Wellesley, MA.

Contributors

Contributors include members of the NE Forests 2100 steering committee and all coauthors to the series of five papers published in the Canadian Journal of Forest Research, 2009, volume 39, issue 2.

Matthew Ayres, Dartmouth College, Hanover, NH

Elizabeth Boyer, Pennsylvania State University, University Park, PA

Nicholas Brazee, University of Massachusetts, Amherst, MA

Lynn M. Christenson, Cary Institute of Ecosystem Studies, Millbrook, NY

Sheila F. Christopher, Buffalo State College, Buffalo, NY

Barry Cooke, Canadian Forest Service, Edmonton, Alberta, Canada

Roger Cox, Canadian Forest Service, Fredericton, New Brunswick, Canada

Marc DeBlois, Gouvernment du Québec, Québec, Québec, Canada

Charlie T. Driscoll, Syracuse University, Syracuse, NY

Joan Ehrenfeld, Rutgers University, New Brunswick, NJ

Ivan. J. Fernandez, University of Maine, Orono, ME

Jeffrey R. Garnas, Dartmouth College, Hanover, NH

Linda E. Green, Appalachian State University, Boone, NC

Peter M. Groffman, Cary Institute of Ecosystem Studies, Millbrook, NY

Jessica Gurevitch, State University of New York, Stony Brook, NY

Robin Harrington, University of Massachusetts, Amherst, MA

Katharine Hayhoe, Texas Tech University, Lubbock, TX

R.T. Holmes, Dartmouth College, Hanover, NH

Daniel Houle. Ministère des Ressources Naturelles et de la Faune du Québec, Québec, Canada

Louis R. Iverson, U.S. Forest Service Northern Research Station, Delaware, OH

J. Kiekbusch, Syracuse University, Syracuse, NY

J.D. Lambert, Vermont Institute of Natural Science, Quechee, VT

Manuel Lerdau, University of Virginia, Boyce, VA

Allison Magill, University of New Hampshire, Durham, NH

S.N. Matthews, U.S. Forest Service Northern Research Station, Delaware, OH

K. P. McFarland, Vermont Institute of Natural Science, Quechee, VT

Kevin J. McGuire, Virginia Polytech, Blacksburg, VA

Myron J. Mitchell, SUNY-College of Environmental Science and Forestry, Syracuse, NY

Scott V. Ollinger, University of New Hampshire, Durham, NH

David Orwig, Harvard Forest, Harvard University, Petersham, MA

Daniel Parry, SUNY-College of Environmental Science and Forestry, Syracuse, NY

Jennifer Pontius, U.S. Forest Service Northern Research Station, Durham, NH

A. Prasad, U.S. Forest Service Northern Research Station, Delaware, OH

Andrew Richardson, Harvard University, Boston, MA

Vikki L. Rodgers, Babson College, Boston, MA

T.S. Sillett, Smithsonian Migratory Bird Center, National Zoological Park, Washington, DC

Erik E. Stange, Dartmouth College, Hanover, MA

Kristina Stinson, Harvard Forest, Harvard University, Petersham, MA

Kathleen A. Theoharides, Defenders of Wildlife, Washington, DC

Mark Watson, New York State Energy Research and Development Authority (NYSERDA), Albany, NY

Norman Willard, U.S. Environmental Protection Agency, Boston, MA

Robert Wick, University of Massachusetts, Amherst, MA

CONTENTS

Boxes

Wolfrun Natural Area, Vermont.
Photo by ©Susan C. Morse, used with permission.

FOREWORD

The climate of northeastern North America has changed markedly over the past 100 years and computer models for the region forecast more change to come. Policy makers, land managers, citizens, and scientists must grapple with what this change means for the future of the region and its forests. This report summarizes five climate science papers published in the Canadian Journal of Forest Research (see Box 1). It provides a picture of the potential effects of climate change and provides a concise scientific overview to inform natural resource management and policy decisions.

This report, "Changing Climate, Changing Forests" grew out of a cross-disciplinary synthesis undertaken by a coalition of 38 U.S. and Canadian scientists as part the larger Northeast Forests (NE Forests) 2100 initiative. NE Forests 2100 was funded by the Northern States Research Cooperative (NSRC) and the New York State Energy Research and Development Authority (NYSERDA), and builds on the efforts of the Northeastern Ecosystem Research Cooperative (NERC). The northeast study region spans seven states of the northeastern United States and five provinces of eastern Canada (Fig. 1). The report traces observed and projected shifts in temperature and precipitation and links them to a web of structural, biogeochemical, and wildlife shifts occurring throughout the region's forests.

Evidence of climate change and its impact on northeastern forests has grown stronger with each passing year. Nonetheless, future changes and effects are uncertain. In this context, how should forest managers, policymakers, nongovernmental organizations, scientists, and concerned citizens act? A forward looking approach to forest management and policy that encompasses a range of possible future conditions is likely to help retain the resilience of forest ecosystems and the critical economic and environmental services they provide. We suggest that stewards of Northeast forests consider taking measures that will draw on forests' climate mitigation potential and aid forests' adaptation to changing conditions.

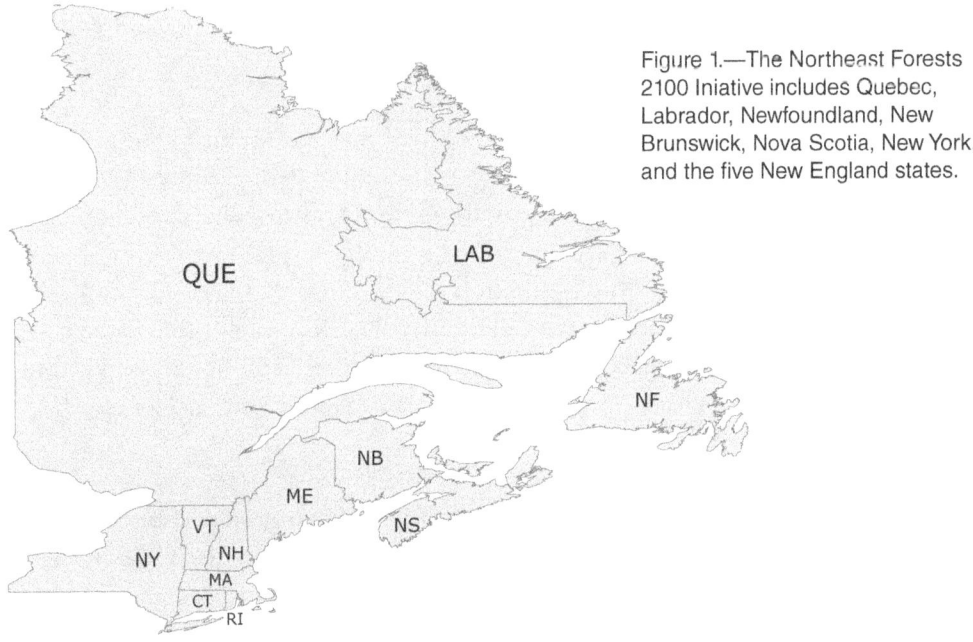

Figure 1.—The Northeast Forests 2100 Iniative includes Quebec, Labrador, Newfoundland, New Brunswick, Nova Scotia, New York, and the five New England states.

INTRODUCTION

What is Climate Change and Why Does it Matter?

Evidence of climate change and its impact on northeast forests has grown stronger with each passing year. Human activities such as fossil fuel combustion, fertilizer production and use, and land use change are driving up concentrations of carbon dioxide and other greenhouse gases in the atmosphere. The emissions of these gases, which are accelerating, are trapping heat and altering the Earth's climate. Our climate shapes our basic living conditions. It controls the growth of agricultural and forest crops that supply our food and fiber, determines our energy needs for heating and cooling, influences the potency of pollutants in our air and water, and drives the melting of glaciers and sea level rise. Even small changes in climate may therefore have major effects on forests and thus for society (Fig. 3).

Climate differs from weather. Weather refers to the day-to-day changes in local conditions such as temperature, precipitation, and humidity. Climate is the long-term average of these indicators across large regions. Because climate is a long-term average, shifts in climate are harder to observe than changes in weather. That's where research comes in. By tracking temperature and precipitation patterns over time and in response to changing atmospheric conditions—such as rising greenhouse gas concentrations –researchers can trace long-term patterns in climate as distinct from the weather patterns we experience day to day.

The Earth's surface air temperature has warmed by approximately 1.4 °F (0.8 °C) over the past century. The amount and patterns of precipitation have also changed (IPCC 2007). Over the next century, temperatures will continue to increase. The increase will depend on the amount of greenhouse gases emitted to the atmosphere from human activities and, to a smaller extent, on natural climate variability.

Model projections of future change based on low and high emission scenarios developed by the Intergovernmental Panel on Climate Change (IPCC) estimate a 5.2 to 9.5 °F (2.9 to 5.3 °C) increase in the Earth's average surface air temperatures by the year 2100 (Nakicenovic et al. 2000).

In 2011, the average global concentration of atmospheric carbon dioxide (CO_2), the most common greenhouse gas, was 387 parts per million (ppm). This is the highest level in the last 800,000 years. Projections for the future suggest that the world will follow or exceed the high emission scenario (Friedlingston et al. 2010).

Rising greenhouse gas concentrations in the Earth's atmosphere have implications for both local and global climates. Over the last century, the patterns of warming in the northeastern United States has been strikingly similar to the patterns for the entire United States and for the whole world (Fig. 2).

The NE Forests 2100 project examined the influence of climate change on the forests of the Northeast. The work, which synthesized historic records, experimental studies, and computer models, reveals important linkages between climate and the basic functioning of the region's forests.

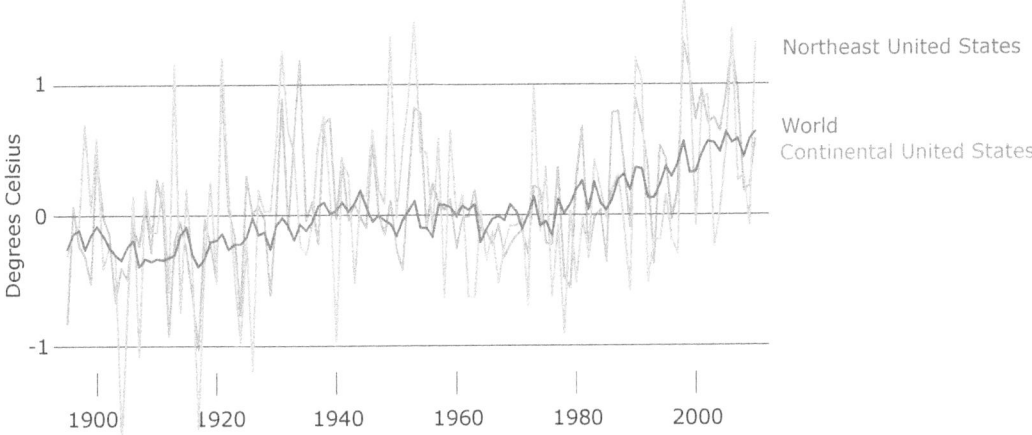

Figure 2.—Mean surface temperature change in the past century: global, continental U.S., northeastern U.S. (compared to the 1951 to 1980 mean). Over the period shown, world temperatures have increased by 1.4 °F (0.8 °C). United States temperatures have increased by 1.5 °F (0.85 °C), and northeastern United States temperatures by 1 °F (0.56 °C). Global data from http://data.giss.nasa.gov/gistemp/tabledata/GLB. Ts+dSST.txt; continental U.S. data from http://www.ncdc.noaa.gov/oa/climate/research/cag3/nt.html; and northeastern U.S. data from http://www.ncdc.noaa.gov/oa/climate/research/cag3/nt.html.

Box 1: Climate Change in the Northeast: Mounting Evidence

This report is drawn from papers published by NE Forests 2100. These, in turn, are built on past research and syntheses. Key references are the following NE Forests 2100 papers, published in Canadian Journal of Forest Research Volume 39, 2009. (Full citations are in the Literature Cited section beginning on page 40).

Climate and hydrological changes in the northeastern United States: recent trends and implications for forested and aquatic ecosystems, by Huntington et al.

Climate change effects on native fauna of northeastern forests, by Rodenhouse et al.

Composition and carbon dynamics of forests in northeastern North America in a future, warmer world, by Mohan et al.

Consequences of climate change for biogeochemical cycling in forests of northeastern North America, by Campbell et al.

Responses of insect pests, pathogens, and invasive plant species to climate change in the forests of northeastern North America: What can we predict?, by Dukes et al.

Other resources:

Karl, T.R., J.M. Melillo, and T.C. Peterson, eds. 2009. Global Climate Change Impacts in the United States. Cambridge University Press. www.globalchange.gov/usimpacts.

Fahey, T.J., F. Carranti, C. Driscoll, et al. 2011. Carbon and Communities: Linking Carbon Science with Public Policy and Resource Management in the Northeastern United States. Hubbard Brook Research Foundation. Science Links Publications, Vol. 1, no. 4.

Frumhoff, P.C., J.J. McCarthy, J.M. Melillo, et al. 2007. Confronting Climate Change in the U.S. Northeast: Science, Impacts, and Solutions. Synthesis report of the Northeast Climate Impacts Assessment (NECIA). Union of Concerned Scientists. 146 p.

NYSERDA ClimAID Team. 2010. Responding to Climate Change in New York State, the synthesis report of the Integrated Assessment for Effective Climate Change Adaptation Strategies in New York State. C. Rosenzweig, W. Solecki, A. DeGaetano, et al., eds. New York State Energy Research and Development Authority (NYSERDA). http://www.nyserda.org/programs/environment/emep/clim-aid-synthesis-draft.pdf

Spierre, S.G. and C. Wake. 2010. Trends in Extreme Precipitation Events in the northeastern United States: 1948-2007. Carbon Solutions New England. University of New Hampshire. http:/www.cleanair-coolplanet.org/cpc/document/2010neprecip.pdf

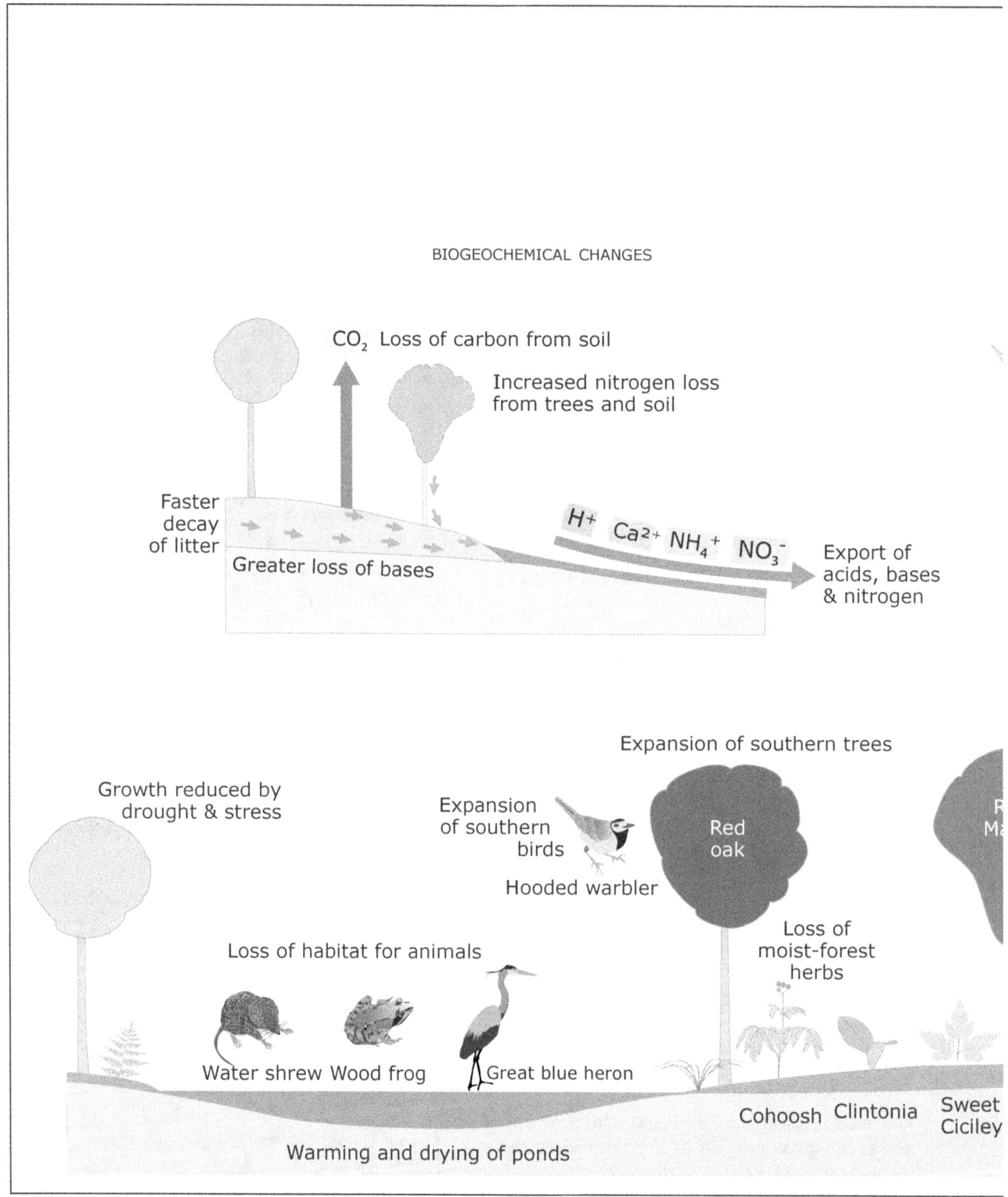

BIOGEOCHEMICAL CHANGES

CO_2 Loss of carbon from soil

Increased nitrogen loss from trees and soil

Faster decay of litter

Greater loss of bases

H^+ Ca^{2+} NH_4^+ NO_3^-

Export of acids, bases & nitrogen

Expansion of southern trees

Growth reduced by drought & stress

Expansion of southern birds

Hooded warbler

Red oak

R Ma

Loss of moist-forest herbs

Loss of habitat for animals

Water shrew Wood frog Great blue heron

Cohoosh Clintonia Sweet Ciciley

Warming and drying of ponds

Figure 3.—Possible climate-driven changes in forests. Original drawing, based on processes discussed in this report. Illustration by Je

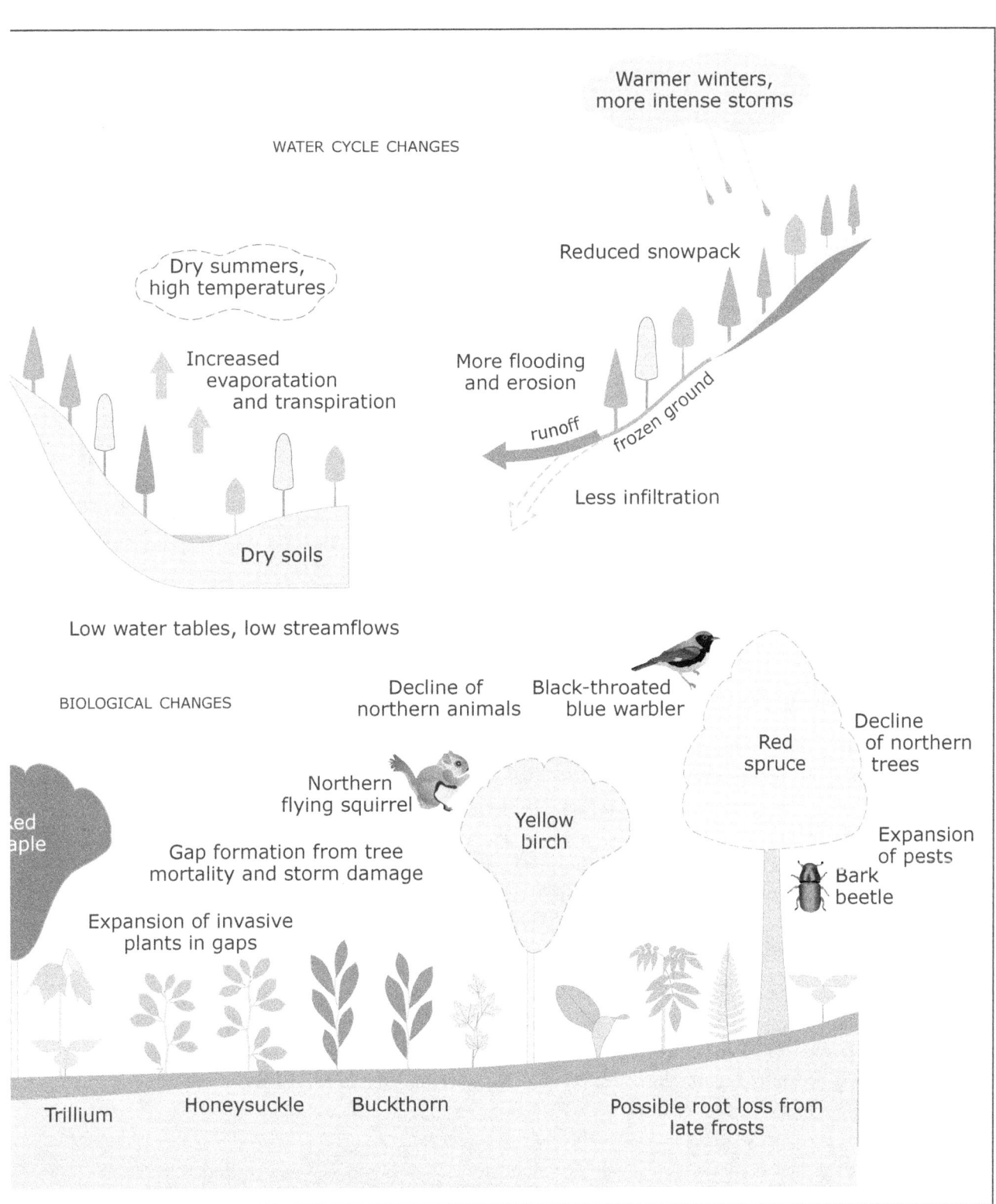

WATER CYCLE CHANGES

Warmer winters,
more intense storms

Dry summers,
high temperatures

Reduced snowpack

Increased
evaporatation
and transpiration

More flooding
and erosion

runoff

frozen ground

Less infiltration

Dry soils

Low water tables, low streamflows

BIOLOGICAL CHANGES

Decline of
northern animals

Black-throated
blue warbler

Decline
of northern
trees

Red
spruce

Northern
flying squirrel

Yellow
birch

Expansion
of pests

Gap formation from tree
mortality and storm damage

Bark
beetle

Red
maple

Expansion of invasive
plants in gaps

Trillium

Honeysuckle

Buckthorn

Possible root loss from
late frosts

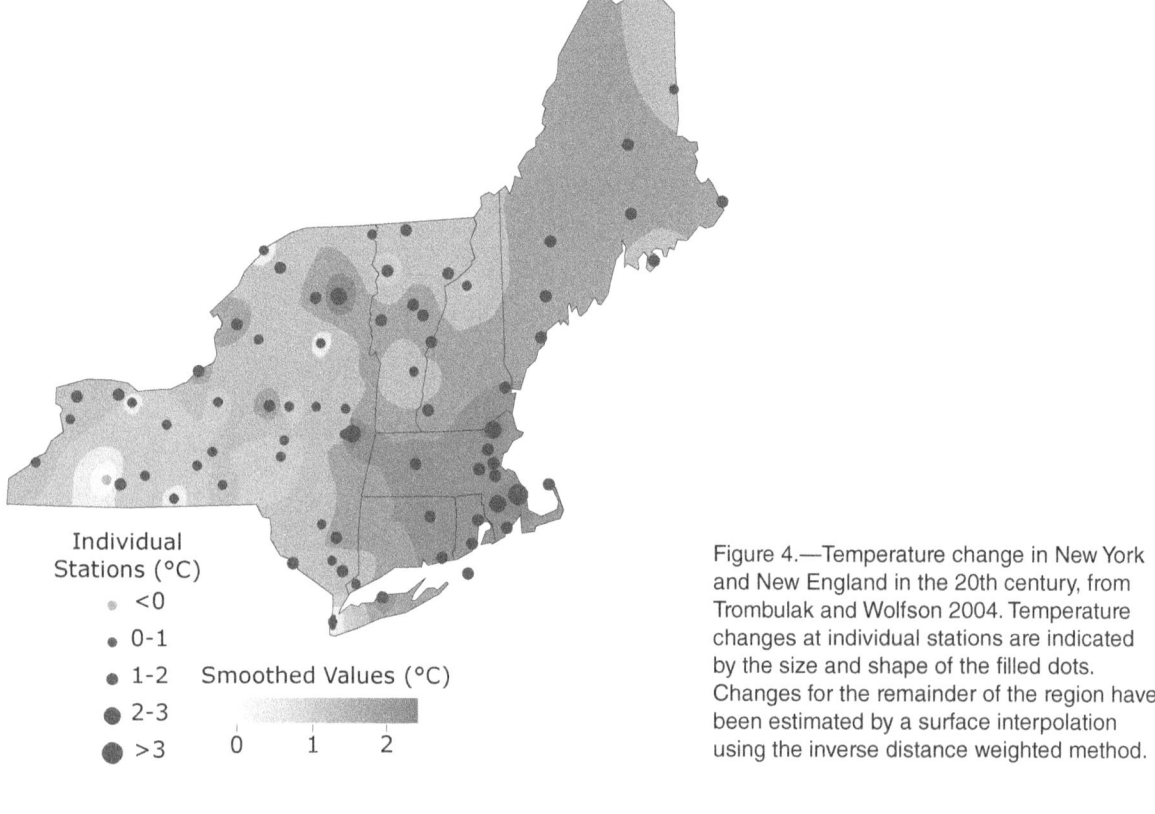

Individual
Stations (°C)
- <0
- 0-1
- 1-2   Smoothed Values (°C)
- 2-3
- >3   0   1   2

Figure 4.—Temperature change in New York and New England in the 20th century, from Trombulak and Wolfson 2004. Temperature changes at individual stations are indicated by the size and shape of the filled dots. Changes for the remainder of the region have been estimated by a surface interpolation using the inverse distance weighted method.

HOW IS THE CLIMATE OF THE NORTHEAST CHANGING?

Summary

Evidence from multiple datasets show unequivocally that climate change is underway in the Northeast, and the rate of change is faster than expected with larger changes observed since 1970. Several long-term datasets suggest that the climate of the region has become warmer and wetter over the past 100 years, and that there are more extreme precipitation events (Hayhoe et al. 2007). Results from regional climate models predict that the Northeast will become even warmer and wetter in the future, but also more prone to drought (Table 1).

Temperature: Observed Change

As part of the NE Forests 2100 project, scientists reviewed long-term datasets for signs of climate change in the Northeast. Analysis of data from 73 meteorological stations showed that surface air temperatures in the Northeast have warmed by an average of 1.44 °F (0.8 °C) over the last century (Fig. 4). The rate of warming is accelerating; surface air temperatures have risen 0.45 °F (0.25 °C) per decade between 1970 and 2000 (Hayhoe et al. 2007).

Average climate conditions tell only part of the story. Some of the most consequential changes in climate involve seasonal shifts and differences in the extreme values of temperature or precipitation. Winter shows the most pronounced warming in the Northeast. The average winter air temperature (December, January, and February) has increased 1.3 °F (0.7 °C) per

Table 1.—Observed and projected changes in the regional climate of the northeastern United States. The results are based on output from running the Parallel Climate Model with a low emissions (B1) scenario and the Hadley Climate Model with a high emissions (A1FI) scenario. See Fig. 5 for details on the scenarios. Adapted from Hayhoe et al. 2007.

Climate Variable	Total Historical Change	Historical Change 1970-2000	Projected Change to 2099, Low Emissions Scenario	Projected Change to 2099, High Emissions Scenario
----- Temperature, °F (°C) -----				
Annual temperature	1900-1999	+1.44 (0.80)	+1.35 (0.75)	+5.2 (2.9)	+9.5 (5.3)
Winter (DJF) temperature, °C	1900-1999	+2.16 (1.20)	+3.78 (2.10)	+3.1 (1.7)	+9.7 (5.4)
Summer (JJA) temperature, °C	1900-1999	+1.26 (0.70)	+0.65 (0.36)	+4.3 (2.4)	+9.0 (5.0)
----- Precipitation, inches (mm) -----				
Annual	1900-1999	+3.94 (100)	-0.94 (24)	+2.84 (72)	+5.67 (144)
Winter	1900-1999	-0.20 (5)	+0.35 (9)	+0.99 (25)	+2.47 (63)
Summer	1900-1999	+0.39 (10)	-0.04 (0.9)	+0.11 (2.8)	0 (0)
----- Phenology, days -----				
First leaf	1916-2003	-3.8	-6.6	-6.7	-15
First bloom	1916-2003	-3.6	-2.8	-6.3	-16
Snow cover per month	1950-1999	-0.2	-1.6	-2.4	-3.8

decade over the past three decades. The average summer temperature (June, July, and August) has risen just 0.2 °F (0.1 °C) per decade for the same period. The winter warming is consistent with the findings of a recent regional analysis of mean, minimum and maximum winter air temperatures, which showed increases ranging from 0.7 to 0.8 °F (0.37 to 0.43 °C) per decade between 1965 and 2005 (Burakowski et al. 2008).

Both minimum and maximum temperatures are increasing, with minimums rising more than maximums. For example, a greater proportion of measurement stations showed warmer minimum temperatures than warmer maximum temperatures from 1960 to 1996 (DeGaetano and Allen 2002). These changes have decreased the diurnal temperature range, which is the temperature fluctuation in a single day. The diurnal temperature range is important to plants: insufficient diurnal temperature range can inhibit flowering, seed production, or germination.

Temperature: Projected Change

A regionally down-scaled model tailored to the Northeast was used to forecast how climate in the region is likely to shift in the future under low and high greenhouse gas emission scenarios (see Box 2). Model projections suggest that by the end of the century the mean annual temperature will increase 5.2 °F (2.9 °C) for the low emission scenario and 9.5 °F (5.3 °C) under the high emission scenario. Seasonally, the model suggests that, contrary to what has been observed over the past 35 years, the temperature increases will be greater in summer than in winter (Hayhoe et al. 2007).

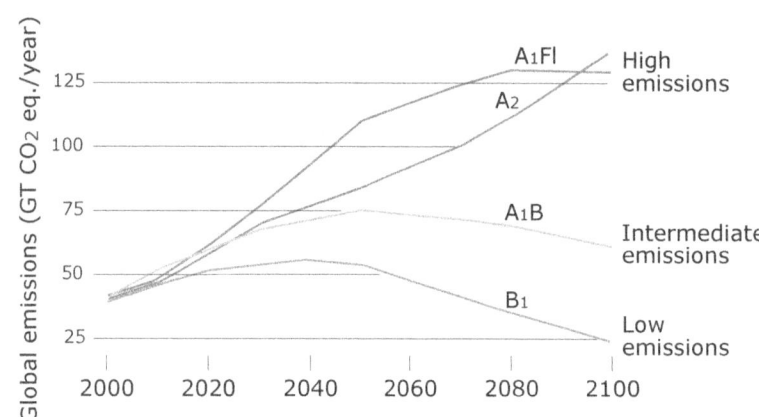

Figure 5.—The Intergovernmental Panel on Climate Change greenhouse gas emissions scenarios from 2000-2100. Emissions are measured in gigatons (billions of tons) of carbon dioxide equivalent, which includes carbon dioxide, methane, nitrous oxide, and fluorine gases. The NE Forests 2100 project used the A1FI high and B1 low emission scenarios. Adapted from IPCC (2007).

Box 2: New Climate Models: Scaling Down from Continents to Regions and Sites

Computer models provide scientists with tools to forecast how changes in greenhouse gas emissions may influence future climate. Climate forecasts can inform policy decisions aimed at mitigating climate change as well as management strategies for adapting to it. The Intergovernmental Panel on Climate Change (IPCC) developed a series of scenarios to describe and quantify how global greenhouse gas emissions may change in the future. These scenarios are based on projected changes in human population size, economic activity, and energy utilization, which exert considerable influence on emissions. NE Forest 2100 scientists used these scenarios in an updated regional climate model to project how climate and other important environmental conditions will change in the future.

IPCC Emissions Scenarios

The IPCC has developed many different emissions scenarios (Fig. 5). NE Forests 2100 used two; a low emission scenario and a high emission scenario. These two scenarios capture the range of possible climate futures, as described below.

High emissions scenario (A1FI)—This is a fossil fuel intensive future with rapid economic growth and a global population that reaches 9 billion in 2050 and then gradually declines. New energy efficiency technologies are not employed until late in the century. Atmospheric CO_2 concentrations rise to 940 ppm, greater than three times the preindustrial concentration, by 2100.

Low emissions scenario (B1)—This is a low fossil-fuel future. Economic growth is principally in the service and information sectors and clean and efficient technologies

are adopted. As with the high emission scenario, global population reaches 9 billion in 2050 and then declines. Atmospheric CO_2 concentrations rise to 550 ppm, double preindustrial concentrations, by 2100.

Northeast Climate Model—Updated Features

The NE Forests 2100 synthesis builds on new modeling techniques to forecast the impact of the high and low IPCC emissions scenarios on the climate of the Northeast. The older models treated all of New England as a single data point, or cell. The improved model uses a technique called statistical downscaling to estimate changes at a much finer scale (Hayhoe et al. 2007). The new model:

Assembles nine coupled atmosphere-ocean general circulation models (AOGCMs) that address important interactions among the major components of the climate system.

Generates climate forecasts for specific locations at a spatial resolution of approximately 6 miles (10 kilometers) rather than hundreds of miles.

Provides temperature and precipitation output at a daily rather than monthly interval.

Provides temperature and precipitation estimates that can be used as input to a detailed hydrologic model called the variable infiltration capacity (VIC) model. The VIC model simulates climate effects on hydrologic variables such as evapotranspiration, runoff, snow water equivalent, and soil moisture which can then be used to model suitable habitat and nutrient cycling.

Projections based on future emissions scenarios indicate growing season length will increase by 29 to 43 days by the end of the 21st century (Hayhoe et al. 2007). The longer growing season will result in a 10- to 14-day advance in the onset of spring and a delay in fall senescence and leaf off. These changes will have a profound impact on the region's forests and water cycle including productivity, plant nutrient uptake, streamflows, and wildlife dynamics.

Precipitation: Observed Change

Long-term records for New England show that the average annual precipitation of 40.8 inches (1040 millimeters [mm]) has increased by about 3.7 inches (95 mm), or 9 percent, over the last century (Huntington et al. 2009). The largest increases were in the spring and fall (Hayhoe et al. 2007). Summer and winter precipitation changed little. The changes were larger near the coast than inland (Keim et al. 2005). In addition to the changes in the amount and seasonality of precipitation, there were changes in the precipitation extremes and the ratio of rainfall to snowfall (Huntington et al. 2009).

Both the intensity and frequency of extreme precipitation events have increased over the past century. This is true whether events are measured by frequencies, by percentiles, or by recurrence intervals. This is shown by a recent analysis of extreme events for 213 weather stations for the period 1948 to 2007 (Spierre and Wake 2010). Using a subset of the weather stations in the Northeast with records dating back to 1900, they show that large precipitation events that occurred historically at a frequency of one per year have increased in frequency by 8 percent (13 events per 12 years).

Despite a trend toward more precipitation, the Northeast is seeing longer periods without rainfall and longer growing seasons. The result is a drier growing season, especially during the summer months, when temperatures and evapotranspiration are high. This summer drying trend is exacerbated by reduced recharge from spring snowmelt. Data from the U.S. Historical Climatology Network indicate that over the last half of the 20th century an increased proportion of winter precipitation has occurred as rain rather than snow. This finding is consistent with observed decreases in snowpack depth at several sites in Maine (Hodgkins and Dudley 2006) and a 9-day reduction in snow-covered days across the region (Burakowski et al. 2008).

Precipitation: Projected Changes

By the end of the century, the average amount of precipitation that falls each year is expected to increase by 7 percent under the low emissions scenario and 14 percent under the high emissions scenario. Precipitation is more difficult to predict than temperature, and these predictions are correspondingly less certain.

On a seasonal basis, precipitation increases are expected to be greatest in winter (12 to 30 percent increase), with much of this precipitation occurring as rain. As a result, the average number of days with snow on the ground in the winter months (December, January, and February) is projected to decrease by as much as 5.2 days per month by the end of the century. Under the high emissions scenario, models suggest that the length of the winter snow season could be cut in half by 2100 in parts of New England (Hayhoe et al. 2007).

Precipitation is not projected to increase in the summer months. However, the intensity of precipitation may increase, consistent with past trends and the expectation that climate warming will lead to an intensification of the hydrologic cycle (Huntington 2006).

Ecological Implications of Changes in Climate

Climate exerts strong influence over ecological functions, such as water use and plant productivity, that have critical impacts on forests. Warmer winters and a longer growing season will increase evaporation and water use by forests. Greater water use will likely reduce summertime soil moisture and increase the occurrence and length of droughts. Drought will decrease forest productivity and increase the susceptibility of trees to insects and disease, with ripple effects on fall foliage, wood supply, and other economic resources. In addition to these direct forest effects, the projected changes in temperature, snowfall, and rainfall will likely prompt a cascade of changes in the water cycle, resulting in altered conditions in the region's rivers and streams. These are considered next.

The average number of days with snow on the ground in the winter months is projected to decrease by as much as 5.2 days per month by the end of the century.

HOW IS THE WATER CYCLE IN THE NORTHEAST RESPONDING TO CLIMATE CHANGE?

Summary

The water cycle of the Northeast is already changing, sometimes in unexpected ways. Long-term data from more than two dozen rivers and dozens of ice monitoring locations show that the region is experiencing earlier snowmelt, earlier spring flows, higher flood flows, and shorter periods of ice cover. Future temperature increases are likely to shift more winter precipitation to rain, leading to higher average winter flows, greater likelihood of ice-jam flooding, reduced summer streamflow, and longer periods of summer drought.

Streamflow: Observed Changes

Climate-driven patterns in precipitation and temperature directly affect the timing and amount of water discharged to rivers and streams. In the Northeast, precipitation is more or less evenly distributed throughout the year, but in some seasons it accumulates in the snow pack, and in others, it is returned to the air by evaporation and transpiration. As a result, streamflows naturally vary greatly by season.

The water cycle responds to changes in the amount and timing of precipitation and to water use by the trees and other plants that cover the landscape. As precipitation increases across the region, more water is available for streamflow. However, greater evapotranspiration by the forests can reduce flow during the ecologically sensitive spring and summer periods.

Historical data for rivers in the Northeast show changes in the amount and timing of flows. Over the last 100 years, <u>average</u> <u>annual</u> streamflow increased at 22 of 27 sites on rivers in New England (Hodgkins and Dudley 2005). In addition, peak flows came earlier. Streamflow data from 11 rural rivers show that high spring flow (as measured by the date on which half of the water discharged from January through May has passed the gage) is occurring 1 to 2 weeks earlier now than in the 1930s (Hodgkins et al. 2003). Average March flows have increased and average May flows have decreased, lowering the May peak and making flows more uniform during the snowmelt season. These changes are consistent with the impact of reductions in the snowpack and warmer late winter temperatures. Hartley and Dingman (1993) reached similar conclusions. They found that maximum river flows in watersheds across the region occurred approximately 5.4 days earlier for each 1.8 °F (1 °C) increase in average annual temperatures. Peak river flows on most of the streams analyzed also increased over the past 75 years.

Streamflow: Projected Changes

Climate models suggest that changes in the water cycle will create even more pronounced shifts in spring runoff and summer low-flow conditions. In the future, winter rainfall is likely to increase, producing higher winter streamflows. With more precipitation falling as rain, the duration and extent of snow cover is projected to diminish, leading to earlier runoff and lower spring peak flows. A consequence of this change is that spring snowmelt flows will drop to summer levels earlier in the season, flattening the streamflow hydrograph and leading to earlier summer drying (Huntington 2003).

Decreased snowpack, increased water use by plants, and shifts in precipitation patterns are all likely to decrease summer flows. Snowpacks typically release their stored water slowly, thereby recharging groundwater and maintaining base flow conditions in the summer months. With more winter rainfall and warmer winters, winter runoff will likely increase, and overall snowpack depth and duration o.f snow cover will decrease Water storage may thus decrease, providing less water for summer periods. The trend towards more intense storms with longer intervening dry periods will increase summer drying. This is already being observed.

The U.S. Fish and Wildlife Service uses the August median flow as an indicator of low flow conditions that are critical to native fish. A recent analysis for the northeastern United States indicates that by 2100, on average, streamflow will arrive at critical levels about 1 week earlier in the summer and stay below these levels up to 3 weeks longer in the fall. This extended low-flow period would reduce the availability of high quality habitat for fish during the summer period.

Ice Cover: Observed & Projected Changes

Ice cover of lakes and rivers offers another signal of the changing climate of the Northeast. At eight monitored lakes in New England, ice-out (the last date with ice) has advanced by 1 to 2 weeks over the past 100 years (Hodgkins et al. 2002). Similarly, ice-out on many rivers in the northeastern United States has advanced and the total number of days with ice cover has decreased substantially since the 1930s (Hodgkins et al. 2005). On one river where ice thickness has been routinely measured since 1912, the thickness on or about 28 February has decreased by 46 percent (Huntington et al. 2003).

Future ice cover changes in lakes and rivers in the Northeast have not been modeled. It is likely that the current trend of shorter ice cover periods and earlier ice-out dates will continue. In addition, higher winter streamflows may increase the frequency of midwinter ice jams and associated flooding (Prowse and Beltaos 2002). Ice-jam flooding in New England can scour riverbanks, damage roads and infrastructure, and have adverse ecological impacts.

Implications of Changes in the Water Cycle

Climate-driven changes in the water cycle may have profound consequences for ecosystem services. High spring flows are necessary for the migration of Atlantic salmon, and are important for hydropower and recreational boating. Increased periods of low flow are stressful for fish and wildlife. Some of the environmental consequences of changes in the water cycle are described below, and the economic consequences are discussed in Box 3.

> *Water Quality* Rain on snow events are expected to become more prevalent in a warmer winter climate and can result in increases in acid pulses to streams in acid-sensitive areas. Acid pulses release inorganic aluminum which is toxic to many aquatic organisms. Increases in streamwater nitrate, associated with warmer winters, may increase the intensity of acid pulses. This has been projected for a sensitive site, Hubbard Brook Experimental Forest in the White Mountains of New Hampshire, under a changing climate (Campbell et al. 2009).

Box 3: Our Climate, Our Selves: Social and Economic Implications of Climate Change in the Northeast

The Northeast's economy is strongly tied to the region's climate. In a region with natural resource-based traditions and industries, limits on resources mean changes and challenges to the landscape and its citizens.

Winter Recreation

Snowmobiling Winter recreation, and snowmobiling in particular, is expected to decline as temperatures rise and snow cover declines. There are about 40,500 miles of snowmobile trails in Maine, Massachusetts, New Hampshire, New York, Pennsylvania, and Vermont, collectively, and these trails account for about a third of the trails in Nation. These trails generate approximately $3 billion in revenue a year (International Snowmobile Manufacturers Association 2006). Fewer snow-cover days pose major challenges to this industry. Other recreational activities dependent on natural snow, such as cross-country skiing and snowshoeing, will also face tough conditions with changing winter precipitation patterns.

Downhill Skiing and Snowboarding The skiing and snowboarding industry is likely to face shorter seasons along with increased snow-making requirements and associated operating costs as the winters get shorter and warmer. It is projected that 8 out of 14 ski areas in the Northeast would require at least 25 percent more machine-made snow in the next several decades (Frumhoff et al. 2007). In some more southerly areas, conditions are projected to become too warm to reliably and efficiently make snow (Frumhoff et al. 2007).

Ice Fishing Later and shorter ice cover periods has had a negative impact on regional ice fishing derbies. Media accounts in recent years lament the cancellation of fishing derbies due to unsafe ice conditions from Massachusetts to upstate New York. Ice fishing is part of a way of life in the region and can bring important income to rural towns.

Water Dependent Resources

Hydroelectricity Hydroelectric power depends largely on snowmelt runoff collected in reservoirs in the spring and released throughout the year to meet the region's energy demands. As winter precipitation patterns shift with a higher proportion occurring as rain, the amount of snowmelt water available for long-term storage is likely to diminish, placing strains on peak power and summer base load production, as demand for energy for cooling is likely to increase.

Public Water Supply Fresh water is an important but finite resource that is often taken for granted in the Northeast.

Much of our drinking water comes from reservoirs and aquifers that are recharged by snowmelt. If snowmelt decreases and summers get dryer, supplies may no longer be able to meet demands.

Fisheries Fishing is culturally and economically important. Freshwater fish in the Northeast face the dual risk of increased water temperatures and lower summer flows as climate continues to change. According to a study by the U.S. Environmental Protection Agency, the rise in water temperatures that may occur by the end of the 21st century could result in a 50 to 100 percent loss of habitat for brown, brook, and rainbow trout—cold-water species that are highly valued by recreational anglers (Michaels et al. 1995).

Forest-based Industries

Forest industries are major contributors to the regional economy. In 2005, the northern New England revenues from forest industry and forest-based tourism were estimated at $19.5 billion (Frumhoff et al. 2007). As climate changes, both the nature and composition of the region's forests and the conditions under which the forest industry operates could undergo major transformations.

Pulp and Paper The pulp and paper industry remains economically important in the region, particularly in Maine where it is a $1.4 billion industry. The industry faces substantial economic challenges from increased global competition, mechanization of operations, and restructuring. The reduction of suitable habitat for the spruce-fir forests under a high emissions scenario and the likely reduced forest productivity would compound the stresses and job losses in this industry.

Maple Syrup New England and New York produce roughly 75 percent of the U.S. supply of syrup, with a value of about $25 million/year (NERA 2001). Sap flow requires the combination of warm days and freezing nights. Climate change is disrupting the pattern. In central New England, the start of sugaring seasons has shifted from mid-March to early February, producing a shorter tapping season and lower grade syrup.

Seasonal Tourism Fall foliage tourism accounts for 20 to 35 percent of annual tourism in Vermont and Maine. The increased frequency and duration of drought, potential shift from colorful sugar maples to muted oaks, and later fall frosts could contribute to a less brilliant fall foliage. A less vibrant display could have significant economic consequences—a 50 percent reduction in fall tourism could account for up to a 1 percent drop in Vermont employment, with smaller impacts in other states (NERA 2001).

Fish Populations Fish species that spawn in the spring may be most vulnerable to shifts in the timing of snowmelt runoff. If the timing of the spring migration of juvenile salmon from freshwater rivers misses the peak spring flows by as much as 2 weeks, salmon survival could decrease (McCormick et al. 1998).

Low Summer Flows Reductions in summer low flows increase stress to aquatic organisms because of reduced available habitat and increased water temperatures. Drier summers could also result in less groundwater to support in-stream flows, cool-water seeps in streams, and drinking water. Decrease in these flows could harm cold water species, such as trout, that rely on these cool-water refuge areas.

Lake Ice Changes in the timing of ice-out may change the likelihood and/or timing of ice-jam flooding, the rate of summer oxygen depletion in lakes, and the productivity and abundance of aquatic organisms.

Wooded stream in Franconia State Park, NH.
Photo by Elizabeth Morin, used with permission.

HOW ARE NORTHEAST FORESTS RESPONDING TO CLIMATE CHANGE?

Summary

It is difficult to directly measure the response of long-lived organisms, such as trees, to changing climate. To address this challenge, scientists forecast changes in habitat conditions and then project how tree species are likely to adjust to these new conditions, or "climatic envelope." This climatic envelope approach can be used to project shifting conditions for both trees and wildlife. Current modeling studies project that the dominant tree species in the region are likely to undergo dramatic range shifts as forests slowly disassemble and reassemble in response to changes in suitable habitat over the next 100 years. Projections suggest that suitable habitat for spruce-fir forests may virtually disappear from the Northeast in the next 100 years, and that habitat for the northern hardwood trees that currently dominate the region is likely to be replaced by conditions better suited to oak forests (Box 4).

How the productivity of forests will change is unclear. Longer growing seasons and higher atmospheric CO_2 concentrations may increase productivity. Increased summer drought, changes in suitable habitat, changes in pests and disease, and continued problems with air pollution, nitrogen deposition, and acid rain may decrease productivity. What is clear is that climate change increases the uncertainty about the future of the region's forests.

Forest Composition: Observed Changes

Northeastern forests of today are dominated by hardwood and coniferous tree species iconic to the region, such as maples, birches, beech, spruce, and fir. But this has not always been so. Pollen and microfossil records from the Northeast reveal that climate has exerted a strong

---

Box 4: A Tough Transition for Future Forests

Forests will undergo a tough transition as the forest types common today give way to new mixes of species in the future. As forests disassemble and reassemble in response to global change, patterns are unlikely to match precisely the projections of suitable habitat developed by computer models due to slow migrations, lag times, the fragmented nature of remaining forests and surprises that nearly always occur. Human-driven climate change is generally faster than what forests have experienced over the past 120,000 years and some species will have difficulty keeping pace. Some of the factors that may limit their ability to track changing habitats are:

Reproduction and Recruitment Climate change may disrupt the critical synchronies between the timing of flowering, pollen availability and seed development of trees, and the life cycles of the animals that pollinate them and disperse their seeds.

Migration Rates Trees migrate slowly. DNA evidence suggests that the tree migrations that occurred after the last glaciations were much slower than what is needed to keep pace with current and projected climate change. It is likely that the movement of trees will lag behind the movement of climate zones where they currently live. How much behind, and whether this will trap them in unsuitable climates that will further slow their migration, is unknown.

Other Factors The actual future ranges of tree species and forest types are complicated by several factors that are not included in the current models. These include change in water tables, disturbance frequencies, nuisance pests and pathogens, and competition with native and exotic plants. For example, the increased abundance of woody vines that has been observed in temperate forests over the last few decades is apparently associated with increases in atmospheric CO_2 (Dukes et al. 2009).The related decline in forest regeneration due to competition with vines makes future forest establishment, composition, and structure even less certain.

Table 2.—The number of common tree species with projected changes in suitable habitat in 2100 under different emissions scenarios. Adapted from Mohan et al. 2009. The results are for the B1 low emissions scenario use the Parallel Climate Model and those for the A1FI high emissions use Hadley Climate Model. See Hayhoe et al. 2007 for details.

Scenario	Species with decreased habitat	Species with habitat unchanged	Species with increased habitat
Low emissions (B1)	26	10	48
High emissions (A1FI)	33	1	50
Ensemble of 8 scenarios	31	6	47

influence on forest composition over time. In the early Holocene, 12,000 years before the present, the climate was cool and spruce and jack pine woodlands were common on recently deglaciated lands. Three thousand years later, as temperatures warmed, white pine, oak, and eastern hemlock expanded their ranges. Three thousand years after that, as the climate became cooler and moister, fire decreased and northern hardwoods such as beech and maple came to dominate the northern region, while oak, white pine, hickory, and birch became more prevalent in the south. Over the last 1,500 to 1,000 years, as the climate continued cool, white pine declined in abundance and boreal trees like spruce and fir moved south.

While there is clear evidence that our forest trees can migrate with changing climate, past rates of migration have been slow. Most studies suggest that the historical rates of species movement are too slow to keep up with current and future climate change. This mismatch between relatively rapid changes in climate and the slower rate at which tree species migrate complicates the picture of how forests might adapt to climate change in the future.

There is only limited evidence of tree species shifts in response to climate change during the past four decades. One study in Vermont showed that the composition of the forests on the western slopes of the Green Mountains had changed between 1964 and 2004, and as a result, the boundary between the northern hardwood forest and boreal forest had shifted upslope by 299 to 390 feet (91 to 119 m) (Beckage et al. 2008). The authors suggested that climate warming contributed to this change. Other scientists have offered alternative explanations including regional trends in land use, acid rain, and the associated depletion of soil calcium. Another study showed that northern tree species of the eastern United States were regenerating more successfully farther north in relation to their associated adult (i.e., seed source) trees (Woodall et al. 2009). This finding suggests that these species are regenerating more successfully in the north than they used to. But, direct evidence of successful range expansion was not found. Ongoing inventorying, monitoring, and analysis will be critical to documenting and understanding future forest change.

Forest Composition: Projected Changes

Scientists forecast the impact of climate change on forest composition using a "climatic envelope" approach (Table 2). This is a three-step process. First, forest inventory data are examined to determine habitat conditions under which specific tree species currently grow. Second, climate models are used to project future habitat conditions. Third, the changes in

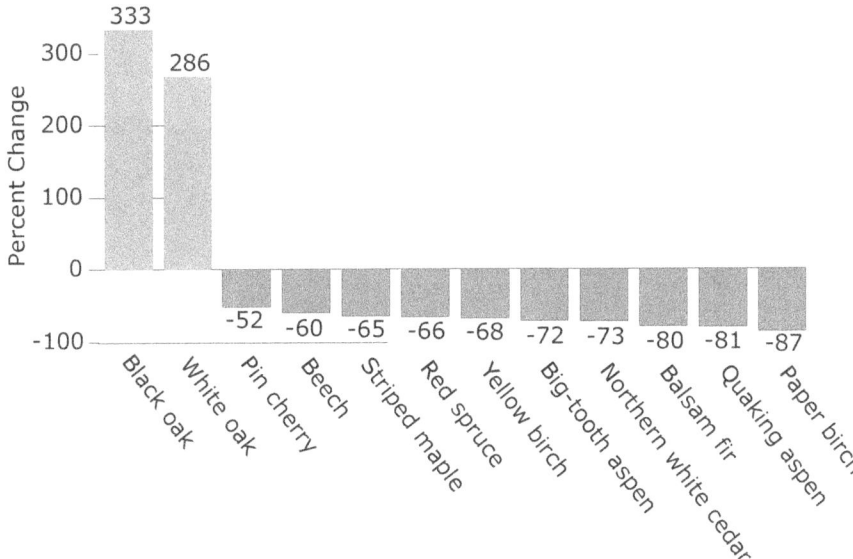

Figure 6.—The 12 tree species showing the largest projected changes in suitable habitat in 2100 under an average-high emissions scenario. Adapted from Mohan et al. 2009.

habitat conditions are used to forecast how the distribution of tree species and the composition of forests may change in the future. The <u>result</u> is typically a map of potential changes in <u>suitable habitat</u> for different tree species and forest types (e.g., spruce-fir, oak-hickory, and maple-beech-birch). Note that, while the climatic envelope approach forecasts the eventual change in tree distributions, it does not forecast how fast these changes will occur. The long life span of trees, the slowness with which they disperse, and the possibility that they may adapt genetically to changed climates all make it unclear how soon, if ever, the trees migrate to the places where the models say they should be. Our current models will have to become considerably more detailed before they can make predictions on the actual distribution of tree species in the future.

Suitable Habitat

Using a suite of four climate models and two emission scenarios together with a vegetation model called the Random Forests Model, scientists predicted changes in suitable habitat for 84 common tree species in the Northeast (Iverson et al. 2008). The model draws on U.S. Forest Service Forest Inventory and Analysis (FIA) data from more than 100,000 plots in the northeastern United States and integrates seven climate variables, 22 soil variables, five topography variables, and four land-use variables. Model results for the 84 tree species showed that suitable habitat is projected to increase for 47 species, decrease for 31 species, and remain unchanged for six species compared to current suitable habitat. The number of species that are projected to gain or lose ground under the two emissions scenarios is shown in Table 2. The results show that many ecologically and economically important species, such as sugar maple and balsam fir, are likely to lose considerable suitable habitat. Oaks will generally do better, with the habitat for some oak species increasing as much as three times (Fig. 6).

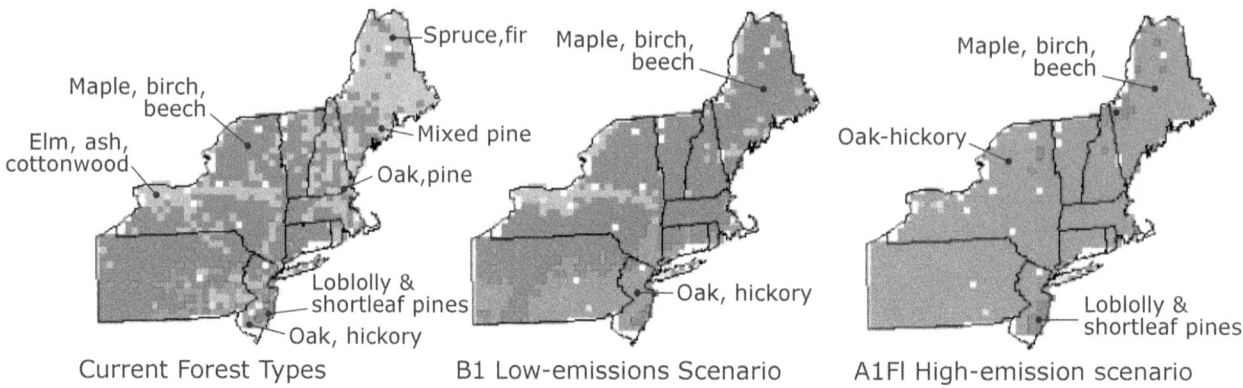

Figure 7. —Current and projected suitable habitat for major forest types in New England under low and high emissions scenarios. See Figure 5 for details of the scenarios. Under the low emissions scenario, the conditions will favor maple-birch-beech forests, while under the high emissions scenario suggest that conditions they will favor oak-hickory forests. Adapted from Iverson et al. 2007.

Forest Types

Maps of projected changes in suitable habitat for several individual species were combined to produce forest habitat type maps for the region under different emissions scenarios. These were then compared to the current distribution of forest types as determined by FIA (Iverson et al. 2008). Results indicate that in the future, only the lowest emissions scenario retains any spruce-fir habitat, and that oak-hickory forest habitat increases significantly at the expense of the maple-beech-birch habitat in all scenarios (Fig. 7).

It is important to emphasize that maps of future forest types reflect only habitat suitability and not necessarily where the trees will actually grow in response to climate change over the next 100 years. Climate change will affect the reproduction, recruitment, migration, and genetic resilience of trees in unpredictable ways (Box 5). Scientists expect that forest change will lag behind suitable habitat change, and there will be significant differences between the modeled and actual future forest composition. For example, while the habitat for oak and hickory forests is projected to increase greatly with climate change, most oaks and hickories presently have difficulty regenerating. Consequently, it is projected that oak and hickories may take centuries to expand their ranges.

Forest Productivity: Observed Changes

Forest productivity is the net growth rate of a forest. It is important because it determines the rates at which forests can produce timber and sequester carbon. In ecology, forest productivity is usually measured as net primary productivity (NPP), the amount of biomass added to an area of forest over time, taking into consideration respiration, maintenance demands, and losses. Historical and paleoecological studies document the strong influence of climate on forest productivity in the Northeast. For example, a 400-year record of black spruce growth near the tree line in eastern Canada shows that modern trees are larger and more productive than similar trees of the same species that grew during colder conditions of the "Little Ice Age" (~1240 - 1850 A.D.) (Vallee and Payette 2004).

Box 5: Forest Composition: Do Species Matter?

Climate-driven changes in forest composition have implications for the economic and ecological future of the Northeast. What are the potential consequences of species shifts on important ecosystem services, such as providing clear water, clean air, or wildlife habitat? Do individual species, or mixes of species, matter as long as there are still trees growing vigorously across the landscape? The function and effects of individual tree species within a forest type are not well understood. The eastern hemlock provides one example of the implications of changes in species composition within a forest. As a result of an introduced pest, the hemlock wooly adelgid, hemlock stands are being replaced by hardwood stands across large areas of the Northeast. This major disturbance provides a glimpse of what the cascade of effects might look like as tree species shift in response to global change.

Hemlock forests of the Northeast lose 50 percent less moisture to the atmosphere through transpiration during summer than do nearby hardwoods stands with similar leaf area. The loss of hemlock and resulting increase in transpiration could lead to diminished flows in small streams and reduced aquatic habitat for salamanders and aquatic invertebrates of conservation concern (Ellison et al. 2005).

Many hardwood tree species allow more sunlight to reach the forest floor than hemlocks. The loss of hemlock and resulting increase in incoming solar radiation could result in higher stream temperatures with detrimental effects on sensitive coldwater species such as brook trout.

Several species of mammals, as well as the colorful Blackburnian warbler, are associated with hemlock forests. The loss of hemlock will significantly diminish the habitat for these animals.

Research on modern forests shows that climate is affecting the timing of forest growth, and thus has the potential to affect productivity. With shorter winters, spring leaf-out is occurring earlier. For example, data from northern hardwood forests at the Hubbard Brook Experimental Forest in New Hampshire indicated significantly earlier spring leaf-out and an increase in green canopy duration of about 10 days over a 47-year period (Richardson et al. 2006). In theory, the longer periods of growth should result in increased productivity. In fact, however, other stressors may limit the extent to which forests take advantage of the extended growing season. Several northern species, both hardwoods and conifers, have had periods of decline or lower productivity over the past 100 years (Mohan et al. 2009). In each case, climate played an important role in the decline, suggesting that future climate change may exacerbate the situation (Table 3).

Forest Productivity: Projected Changes

Model projections suggest that forest productivity for individual hardwood species is likely to be enhanced in the future by warmer temperatures and increased concentrations of carbon dioxide (CO_2) in the atmosphere. However, it is not clear whether these modeled gains will be realized across the landscape and/or whether they can be sustained. Other stresses, particularly altered winter freeze-thaw cycles, increased drought and fire potential, air pollution, and heightened vulnerability to pests and disease, can reduce productivity. These stresses are difficult to fully capture in forest models. In the case of spruce-fir forests, models predict a decline under both low and high emission scenarios. The effects of additional stressors are likely to make the decline worse.

Scientists have used a model called PnET-CN to predict changes in NPP over the next 100 years (Ollinger et al. 2008). Results suggest that the productivity of deciduous forests will increase by 52 percent to 250 percent by 2100, depending on the global change model and CO_2 emissions scenario used. Results further suggest that growth enhancements associated with CO_2 fertilization may be nearly equal to or greater than the effects of climate change (i.e.,

Table 3.—Tree decline and associated climate factors. A review of decline episodes for five different tree species in the Northeast indicates that there have been important associations with changes in climate and weather-related conditions which may be further exacerbated as climate changes in the future. Adapted from Mohan 2009.

Species/ Group	History	Role of Climate	Other Factors	References
Birch	Widespread declines since 1944	Maps of birch decline areas coincide with areas of experiencing extended winter thaw cycles	None	Balch 1944 Bourque et al. 2005 Braathe 1995
Sugar maple	26 widespread decline episodes between 1912 and 1986	Prolonged thaw-freeze events and associated fine root damage have been implicated in sugar maple decline	Insects, disease, loss of soil nutrients	Millers et al 1989 Bertrand et al. 1994 Decker et al. 2003 Fitzhugh et al. 2003
Oak	Large areas of oak mortality recorded in New England and the Appalachian Mountains in the early 1900s	Drought stresses have been reported as important initiating factors in oak decline.	Insects, secondary pathogens	Millers et al. 1989
Ash	Widespread dieback in the Northeast since 1920	Drought and freezing damage have been identified as inciting factors, with drought playing a particularly important role	Phytoplasmal disease, Asian beetle, emerald ash borer	Millers et al. 1989 Poland and McCullough 2006
Red spruce	Widespread decline through the Northeast after 1960, increasing over the last few decades	Reduced cold tolerance leads to winter injury which is intensified rapid rates of thaw and subsequent exposure to refreezing	Acid deposition, anomalous weather	Friedland et al. 1984 Johnson 1992 Schaberg and DeHayes 2000 Bourque et al. 2005

extended growing seasons, warmer temperatures) alone. The results for spruce-fir forests are different. The model suggests that the productivity of spruce-fir forests will decline, regardless of the CO_2 effect (Ollinger et al. 2008).

An application of the PnET-CN model to the Hubbard Brook Experimental Forest shows similar results (Box 6). Because the potential fertilizing effects of CO_2 were not incorporated into the model when this study was done, the productivity gains are less.

Confounding Factors

While extended growing seasons and the potential fertilization effect of CO_2 should enhance forest productivity over the next century, there are several major confounding factors that could work against this beneficial response. These include the following:

Competition Competition between tree species is likely to increase as suitable habitat shifts to favor more southern species and some forest types begin to replace others. Most models calculate productivity for "current-day" forest types and do not account for the impact of competition as forest types disassemble and reassemble.

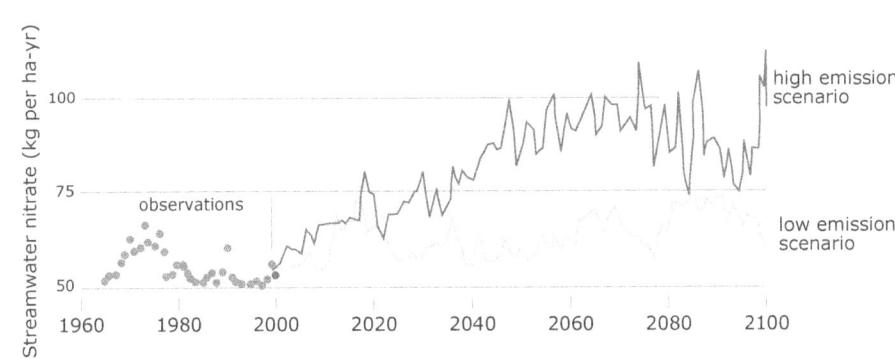

Figure 8.—Measured and predicted nitrate export in streamwater at the Hubbard Brook Experimental Forest The leaching of nitrate from forest soils to streams is projected to increase markedly under a high emissions scenario. Adapted from Campbell et al. 2009.

Box 6 Hubbard Brook Experimental Forest— Climate Change Case Study

To evaluate how forest ecosystems in the Northeast could respond to climate change, downscaled climate projections were developed for the Hubbard Brook Experimental Forest using two climate models, each run under high-emission and low-emission scenarios (550 and 970 ppmv, respectively). The models projected temperature increases of 3.2 to 14.8 °F (1.8 to 8.2 °C) by the end of the century, consistent with trends for the broader Northeast region. Substantial increases in annual precipitation occurred under all scenarios, with increases generally greater during winter months. Streamflows did not change much because increases in precipitation were offset by increases in transpiration.

The climate projections were then used in the PnET-BGC forest ecosystem model to predict changes in forest productivity, nutrient dynamics, and water quality. As in regional analyses, the combination of higher air temperatures and greater precipitation resulted in a longer growing season and increased forest productivity. The estimated productivity gains at Hubbard Brook Experimental Forest were less than in the regional analysis because the benefits of potential CO_2 fertilization were not taken into account. The overall results for 1999 to 2099 was an increase in net primary productivity (NPP) of 8 percent under the low emission scenario, and of 15 percent under the high emission scenario. Wood production increased by 34 percent to 70 percent, dominating the NPP response. Root and foliage production declined slightly.

The water quality of streams that drain forested watersheds reflects the upslope nutrient cycling. The Hubbard Brook model predicted increases of 16 percent to 34 percent in nitrogen mineralization. These increased the nitrification rates and doubled the release of nitrate to streamwater (Fig. 8). The increased nitrate release would convert the Hubbard Brook forest ecosystem from one that retains nitrogen to one that exhibits nitrogen loss. It may also make streams even more acidic and increase the export of nitrogen to coastal waters where it could contribute to eutrophication.

Nuisance Species Climate change in the Northeast is likely to lead to increased exposure and susceptibility to invasive species, pests, and pathogens. Invasive species may have a negative effect on plant productivity if individual trees face increased competition from nonnative plants and woody vines. Pests and pathogens, such as the hemlock adelgid or pine blister rust, are already having major effects on forest productivity in the United States and Canada, and climate change is likely to increase their impact. (For more on nuisance species, see page 32).

Drought Increased evapotranspiration and decreased soil moisture are likely to exacerbate summertime drying and contribute to drought-induced plant stress and decreases in productivity and survival.

Air Pollution Air-borne pollutants can change productivity and influence how trees respond to climate change. Some of the most important pollutants in the Northeast include ground-level ozone, acids, and nitrogen compounds.

Ozone Ground-level ozone can damage plant tissue and decrease photosynthesis. Studies suggest that ozone damage can offset CO_2-induced gains in productivity and make trees more vulnerable to other stresses (e.g., McLaughlin et al. 2007). Ozone levels, which are already high in the Northeast, may increase with climate change as plants produce more volatile organic compounds, which then react with nitrogen oxides to produce ozone.

Acid Deposition Acid deposition already occurs across the Northeast and may increase, especially in high elevation forests, if climate change produces more cloud cover and precipitation. Acidic deposition can impair nutrient availability, reduce reproductive success and frost hardiness, cause physical damage to leaf surfaces, and increase susceptibility to decline.

Nitrogen Deposition Nitrogen is an essential nutrient, but too much nitrogen can mobilize acids and damage forests. Just what the potential impacts of elevated nitrogen deposition in a changing climate will be remain unclear. Some research suggests that nitrogen deposition could help offset natural nitrogen limitations that will persist in the future. Other studies suggest that these limitations will not be important because rising CO_2 may allow increased plant nitrogen uptake and increased nitrogen-use efficiency.

HOW IS BIOGEOCHEMICAL CYCLING IN NORTHEAST FORESTS CHANGING WITH CLIMATE CHANGE?

Summary

Biogeochemical cycling or "nutrient cycling" refers to the movement of elements through soils, plants, surface waters, and the atmosphere. Evidence suggests that climate change will alter biogeochemical cycling in Northeast forests with potentially profound effects on forest productivity, water quality, and other ecosystem services.

Nutrient Cycling: Observed Changes

Climate change can affect nutrient cycling directly though its impact on temperature and precipitation, or indirectly through its impacts on forest composition, growing season length, and the water cycle. These effects can occur on time scales ranging from minutes to millennia. Changes in microbial activity, for example, are nearly instantaneous. Changes in forest composition can take centuries, and changes in soils may take millennia. Figure 3 depicts these effects and highlights feedbacks that may further influence climate change effects on forest ecosystems at different time scales.

Several recent studies show how climate change may alter the cycling of forest nutrients (reviewed in Campbell et al. 2009). These studies suggest that, as climate warms, greenhouse gases will be released from soils, the availability of important nutrients will change, and the water quality in sensitive watersheds will decrease.

Release of Greenhouse Gases from Warmer Soils

Climate-induced changes in nutrient cycling can enhance the release of heat-trapping greenhouse gases from soils and thus accelerate climate warming. These changes are potentially important. Soils contain large and dynamic pools of carbon, and even small changes in these large pools could generate substantial feedbacks to climate warming.

One way this could happen is if the warming of soil increases rates of soil respiration, which is the combined respiration of plant roots and soil microbes. Soil respiration is a key ecosystem process that releases CO_2 from the soil, and hence a key component of the global carbon budget. On the global scale, soil respiration is the second largest component of the global carbon budget, trailing only photosynthesis, and thus even a small increase in soil respiration could equal or exceed the amount of carbon released each year by land use change and fossil fuel combustion combined (Rustad et al. 2001).

Soil respiration rates are strongly influenced by temperature and moisture. Soil warming experiments from Howland Forest (ME), Huntington Forest (NY), and Harvard Forest (MA) showed at least short-term increases in soil respiration with a 9 °F (5 °C) rise in soil temperature (Rustad et al. 1996). Similar results have been reported for 8 out of 14 other ecosystem warming studies surveyed (Rustad et al. 2001). Taken together, these studies provide strong evidence that an increase in temperature will measurably increase the release of carbon from the soil through soil respiration, at least in the short term.

In addition to effects on CO_2, climate-induced changes in soil dynamics could affect the release of other greenhouse gases such as methane (CH_4) and nitrous oxide (N_2O) that are less abundant but more potent than carbon dioxide. Depending on how wet they are, soils can be either a source or sink for CH_4. Most wetland soils release CH_4 to the atmosphere and most well drained forest soils absorb it. In the three soil-warming experiments mentioned above, which were in well drained soils, the rates of CH_4 uptake changed little. In wetland soils, which account for approximately 24 percent of global CH_4 emissions (IPCC 2007), field observations show that warming generally increases CH_4 emissions (Chapman and Thurlow 1996, Christensen et al. 2003).

Soils are the primary source of N_2O to the atmosphere, and warming could potentially influence its emission rate. The evidence, however, is mixed. In the warming experiments at the Harvard and Huntington Forests, increases in temperature alone had no significant effect on the release of N_2O from forest soils (McHale et al. 1998, Peterjohn et al. 1994). However, in another experiment at the Huntington Forest, a combination of warmer _and_ wetter conditions resulted in much higher rates of N_2O loss in heated plots compared to control plots (McHale et al. 1998). In a different kind of climate change experiment at Hubbard Brook, Groffman et al. (2008) experimentally reduced snow pack to simulate future conditions. The loss of snow as insulation, paradoxically, resulted in colder soils, induced mild soil freezing, and resulted in significantly higher rates of N_2O release the following summer (Groffman et al. 2006). The authors attributed this increase to the physical disruption of the soil ecosystem rather than the stimulation of activity by soil microbes.

Changes in Nutrient Availability

The productivity and integrity of forest ecosystems is linked to the supply of nutrients. Climate change can influence nutrient dynamics by altering the rate of litter decomposition, the leaching of key nutrients from the soil, and the uptake of nutrients and water by fine roots and symbiotic fungi.

> _Litter Decomposition_ The decomposition of leaf litter contributes to the formation of soil organic matter and the release of vital nutrients that can be taken up by plants. The effects of warming on rates of litter decomposition were studied in two soil-warming experiments in the northeastern United States: one in a low-elevation spruce-fir stand at the Howland Forest, ME (Rustad and Fernandez 1998) and the other in an even-age mixed-hardwood stand at the Huntington Forest near Newcomb, NY (McHale et al. 1998). These and other experimental soil-warming studies (Rustad et al. 1996) suggest that a 5 °F to 9 °F (3 to 5 °C) increase in mean annual soil temperature can increase decomposition rates for most hardwood litter and, to a lesser extent, the decomposition rate of red spruce litter. The released nutrients can be taken up and recycled by forest vegetation, or leached from soils to streams and lakes.

> _Leaching of Nitrogen from Soils_ Nitrogen is a basic element that is fundamental to the growth of plants. In the Northeast, nitrogen limits forest growth under most conditions. Too much nitrogen, however, can have detrimental effects on soil, trees,

and surface waters. Taken together, nitrogen pollution and climate-induced changes in nitrogen cycling have the potential to cause profound shifts in nitrogen dynamics of Northeast forests. Evidence from both empirical studies and simulation modeling suggests that both the faster organic matter decomposition in warmer soils and the more frequent soil freezing events associated with reduced snow coverage can accelerate nitrogen losses from Northeast forests. One way this may happen is by effects on the process of "nitrification", or the production of nitrate by soil microbes. Once nitrogen is converted to nitrate, it is subject to leaching from soils to surface waters. This acidifies the soil and enriches the receiving surface waters in streams and lakes. Several soil-warming experiments across the region showed a general increase in nitrification of up to 50 percent with increasing temperature (Rustad et al. 2001). Soil freeze events, which are projected to become more common with less snow coverage, have also been linked to increases soil nitrate leaching (Fitzhugh et al. 2001). This process can result in spring pulses of nitrate in stream water (Likens and Bormann 1995).

Leaching of Base Cations from Soils Base cations, particularly calcium and magnesium, are important nutrients that help buffer acidic inputs and support forest growth. The supply of soil calcium is particularly important in northeastern North America because acidic deposition can deplete exchangeable soil calcium and other base cations from forest soils (Fernandez et al. 2003). Climate warming and increased rainfall has the potential to accelerate the rate of base cation loss from soils. Warming-induced longer growing seasons, especially combined with higher growth rates and the potential shift from conifers to hardwoods, may increase the annual calcium and magnesium uptake by forest vegetation. The increasing amount and intensity of precipitation plus the potential increases in the infiltration of water into the soil in winter may result in increasing rates of soil leaching and the cumulative loss of soil calcium and magnesium. Model results indicate that increases in base cation leaching are linked with elevated nitrate leaching (Campbell et al. 2009).

Plant Uptake of Nutrients and Water by Roots Trees have extensive networks of fine roots which provide a large surface area for water and nutrient exchange with the soil. Many symbiotic fungi, including many common woodland mushrooms, are associated with roots. Because of their central role in water and nutrient cycling, fine roots and their associated fungi will play a pivotal role in determining how forests respond to climate change. Evidence from several experimental manipulations suggests that climate change will alter the dynamics of fine roots (Arft et al. 1999, Burton et al. 1998). Palatova (2002), for example, showed that a 60 percent reduction in precipitation plus the addition of nitrogen, at the rate of 89 lbs of nitrogen per acre per year (100 kg nitrogen per hectare per year), resulted in a 30 percent decline in fine root biomass after 2 years of treatment in a Scots pine (*Pinus sylvestris* L.) stand. An experiment at Hubbard Brook suggested that soil freezing associated with reduced winter snow cover can kill fine roots (Tierney et al. 2001). Decreased growth and increased mortality of fine roots can reduce nutrient uptake and cause elevated leaching of nitrogen and phosphorous to surface waters (Fitzhugh et al. 2001).

Change in Water Quality

The water quality of streams that drain forested watersheds reflects the chemical impacts of upslope changes in nutrient cycling. As outlined in the case study in Box 6, simulations for Hubbard Brook indicated that climate change over the next century could double the release of nitrate into streamwater. This could make streams more acidic and degrade water quality.

Main branch of Hubbard Brook. Photo by Hubbard Brook Research Foundation.

HOW IS WILDLIFE IN NORTHEAST FORESTS RESPONDING TO CLIMATE CHANGE?

Summary

Climate affects wildlife through changes in the quality and distribution of habitat, the availability of food, the abundance of parasites and diseases, and the incidence of stress from heat and drought. Ecological specialists and animals whose populations are already declining due to other stressors will be most vulnerable. Species with restricted ranges, species restricted to a single habitat, and species with small isolated populations will be particularly at risk, and are most likely to be affected by the smallest amount of change. In recent decades, climate change has already affected the distribution and abundance of many species. For example, detailed historical information indicates that the ranges of many bird species are already changing and that there will be substantial gains and losses in the future, predominantly among migratory bird species, under both high and low emissions scenarios.

Native Wildlife

Climate affects the native wildlife of forests in the Northeast at all levels of organization, from the physiology of individual animals to changes at the population level. Given the wide range of potential impacts, scientists often focus on specific taxa such as mammals, amphibians, insects, and birds. Our best evidence of climate change impacts on wildlife comes from long-term studies of birds, and we focus on them in this section. Potential effects on mammals, amphibians, and insects will also be discussed.

Birds: Observed Changes

Birds, more than any other taxonomic group, have been the focus of climate change research in the Northeast. These studies draw from decades of bird surveys and indicate there have been measureable changes in the timing of key events such as migration, the distribution and abundance of species, and the amount and quality of habitat for forest birds in the Northeast.

Timing of Key Events and Abundance Migratory birds are arriving earlier and breeding earlier in response to recent climate change (e.g., Waite and Strickland 2006). It is unknown how this will affect reproduction rates and survival and therefore the overall size of these bird populations.

Abundance In addition to changes in timing, many bird species have recently increased or decreased their abundance. Among resident birds (i.e., those that remain in the Northeast year round), 15 of 25 species that were studied are increasing in abundance, which might be expected if abundance was limited by winter climate. Five of the ten remaining species (including many highly valued species such as ruffed grouse) are declining in abundance, and the other five show no detectable trends. In contrast to the residents, the short-distance migrants and neotropical migrants show no overall trends in abundance: the number of increasing species is roughly equal to the number of decreasing species in each group.

Table 4.—Number of bird species projected to change their abundance and range between 2000 and 2100. Emissions scenarios as in Fig. 5. Adapted from Rodenhouse et al. 2008.

Scenario	Abundance declining	Abundance unchanged	Abundance increasing	Range declining	Range unchanged	Range increasing
Low emissions (B1)	60	22	68	33	60	57
Average-high (A2)	56	27	67	32	62	56
High emissions (A1FI)	38	48	48	15	94	41

Range expansion Northward range expansions are occurring in many species. For example, 27 of 38 bird species of northeastern forests for which historical data exist have expanded their ranges predominantly in a northward direction. Of the 27 species expanding northward, 15 are neotropical migrants, six are short-distance migrants, and four are residents. These northward migrations are consistent with climate change, and in fact correlate well with regional climate (Rahbeck et al. 2007). They could also be caused by nonclimatic factors such as changing land use and forest cover, winter bird feeding, or provision of nest boxes. However, most of these nonclimatic factors are occurring throughout the region and therefore cannot explain the strong pattern of predominantly northward range shifts. Climate change thus appears the most convincing explanation for the observed shifts.

Habitat change Birds that depend on high elevation spruce-fir forests and cooler temperatures are uniquely susceptible to climate change. High elevation spruce-fir habitat covers less than 1 percent of the Northeast's landscape and most bird species breeding in this habitat are listed as of conservation concern (Rodenhouse et al. 2008).

Birds: Projected Changes

Climate exerts both direct and indirect effects on birds. Direct effects, such as late spring storms, may kill migrating birds. Indirect effects include changes in habitat quality from alterations in food supplies or shifts in vegetation composition and structure. Statistical models can predict how the distribution and abundance of bird species might shift in response to changing climate and habitat. As with the tree models described earlier, these are climatic envelope models. The models first develop associations between the abundance of each species of bird to habitat variables such as climate, elevation, and the abundance of specific tree species (Matthews et al. 2004). Climate models are then used to predict how the climate will change under a specific emission scenario, and envelope models are used to predict how the tree species will change. The changed climate and habitat variables are then used to calculate how abundance of each bird species will change. The results of climatic envelope modeling for the Northeast indicate major changes in bird distributions and abundances (Rodenhouse et al. 2008). The warmer the projected climate, the greater the change.

The changes include both increases and decreases (Table 4). The decreases involved both decreases in abundance and decreases in the area in which the species is found; typically the changes in abundance are greater than the changes in area. This is consistent with the

Table 5—Projected changes in bird abundance by 2100 by migratory habit. Emissions scenarios as in Fig. 5. Adapted from Rodenhouse et al. 2008.

Scenario	Neotropical Migrants			Temperate Migrants			Residents		
	Declining	Stable	Increasing	Declining	Stable	Increasing	Declining	Stable	Increasing
Low emissions (B1)	15	27	21	17	21	3	33	60	57
High emissions (A1FI)	30	8	25	21	11	9	32	62	56

understanding that although most bird species can tolerate a relatively wide range in climate, elevation, and vegetative composition, they reach their peak abundance only in a narrow range of habitats.

Projected shifts in abundance vary among bird species of different migratory habits (Table 5). For resident species, twice as many are expected to increase in abundance as decrease. For short-distance migrants, about twice as many are projected to decline in abundance as expand. Among neotropical migrants, an equal number are expected to increase as decrease.

These broad trends do not tell the whole story, however. The models suggest that many common and culturally important species will decrease. These include the ruffed grouse and black-capped chickadee among the residents; the iconic Baltimore oriole and hermit thrush among the short-distance migrants; and the colorful Blackburnian warbler and rose-breasted grosbeak and mellifluous wood thrush and veery among the long distant migrants.

The predicted changes vary with geography (Fig. 9; Rodenhouse et al. 2008). Maine and New Hampshire have sufficient areas of suitable habitat and may see large increases in bird species richness (a measure of the diversity of the community). In contrast, the southern portion of the Northeast, particularly Pennsylvania and western New York, may see large decreases in species richness.

Habitat also matters. Habitats with many ecological specialists, such as wetlands and high elevation forests, are likely to have significant losses of their characteristic species (Rodenhouse et al. 2008, 2009). For example, the models suggest that more than 50 percent of wetland bird species, including the common loon and American bittern, could become less common in the northeast as a result of habitat changes (Rodenhouse et al. 2000). Birds that breed in high elevation spruce-fir forests are uniquely susceptible to climate change because of the limited opportunity to shift to new locations. This habitat type covers less than 1 percent of the Northeast and is projected to essentially disappear from the region by the end of the century under the high emissions scenario. The majority of bird species breeding in this habitat are presently listed in one or more of the northeastern states and Canadian provinces as in need of conservation. In the case of the Bicknell's thrush, the most intensively studied high elevation forest bird in the Northeast, a geographic information system (GIS) model (Lambert et al. 2005) projects the loss of half the suitable habitat available to Bicknell's thrush with even a 1.8 °F (1 °C) change in mean annual temperature (Rodenhouse et al. 2008).

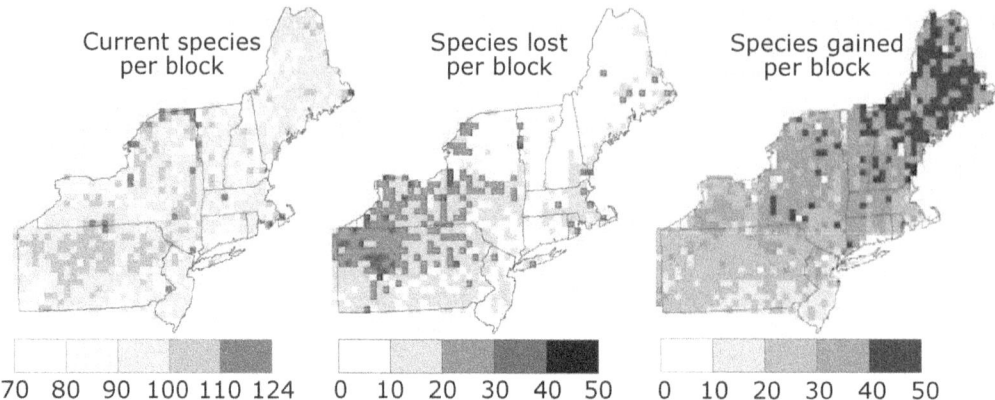

Current species per block	Species lost per block	Species gained per block
70 80 90 100 110 124	0 10 20 30 40 50	0 10 20 30 40 50

Figure 9.—Projected gains and losses in bird species richness across the Northeast under high emissions scenario. Each 20 km by 20 km grid square will lose some species (in red) and gain others (in green). Maine and northern New Hampshire are projected to show a net increase in species richness as range restrictions lift under a warming climate. The southwestern part of the region is projected to show a substantial net loss in species richness as northern species decline. Results are shown for the A1FI scenario, Fig. 5. Adapted from Rodenhouse et al. 2008.

Mammals

Climate change affects mammals through direct thermal stress, shifts in habitat and food availability, increases in parasites and diseases, and responses to extreme weather events. Two species, little brown bat and moose, provide examples of animals that may be impacted by climate change. The little brown bat is vulnerable to changes in food resources and hibernation conditions, whereas moose will be challenged by heat stress, vegetation change, and parasites.

Little brown bat Small mammals with high energy demands, such as bats, may be particularly vulnerable to changes in the food supply. Small bats of the Northeast feed almost exclusively on flying insects, and especially on insects with aquatic larval stages. Climate change may influence the availability of these insects by altering precipitation, stream flow, and soil moisture. The availability of food influences bat reproduction and survival, particularly during hibernation. Bats do not feed during hibernation, and must store enough energy to survive until the insects emerge in the spring. The optimum hibernation temperature is 37 °F (2 °C). Winter temperatures higher or lower than the optimum cause a sharp increase in energy use. It is possible that, as winters warm, bats will experience less stress as their hibernation periods shortens. But, since periods of arousal deplete energy stores, the number of times bats are aroused influence their survival. If little brown bats are aroused more often in warmer winters, they may not be able to store enough energy to survive (Humphries et al. 2004).

Moose Large mammals such as moose may be affected by climate change in different ways than small mammals such as bats. Moose are well adapted to cold temperatures and intolerant of heat. This is true both in summer and winter. Their respiration rates (and energy demands) increase when temperatures exceed 57 °F (14 °C) in summer and 23 °F (-5.1 °C) in winter (Renecker and Hudson 1986). In addition, during hot summers they reduce food intake and can lose body weight. Warming temperatures could shift the lower latitudinal range limit of moose northwards, excluding moose from southern areas of the Northeast. In addition, reductions in snow depth associated with winter warming may bring moose into more contact with white-tailed deer which carry a brain parasite (a meningeal parasite) that can be lethal to moose. Deer avoid areas with permanent heavy snow covers, which were the traditional habitat of moose. As snow depth declines with climate change, moose and deer habitat may increasingly overlap, and more moose may die from the brain parasite.

Amphibians

All amphibians in the Northeast require moist habitats, and so all are potentially sensitive to the changes in temperature and precipitation associated with climate change. One study suggests that amphibians are already responding, with some species calling 10 to 13 days earlier than they were at the beginning of the 20th century (Gibbs and Breisch 2001).

Because most amphibians of the Northeast breed in water, the habitat factor most critical to amphibians is the hydroperiod (the period of time that there is standing water) of ephemeral ponds. The increase in evaporation and frequency of drought associated with climate warming can significantly shorten the hydroperiods and reduce the volumes of ponds (Brooks 2004). The shortened hydroperiod can increase competition, decrease size at metamorphosis, and kill larvae as ponds dry out.

Warmer and dryer climates will impact amphibians in other ways. Negative impacts are likely. For example, terrestrial salamanders may experience increased mortality from decreased moisture levels and reduced effectiveness of anti-predator tactics, and stream-dwelling salamanders may suffer from decreased stream flow and lowered soil moisture. But positive effects are also possible. Warmer winter temperatures may improve the overwinter survival of some species. While it is clear that some local amphibian populations in the Northeast are responding to climate change, more research will be essential to assess the overall risk that climate change poses to amphibian populations across the region.

Insects

Insects are important to overall biological diversity and comprise the base of many food webs. Climate change has been linked to shifts in insect ranges through expansion into new areas and extirpation from areas where they formerly existed. This has been best documented for butterflies. For example, of 35 species of nonmigratory European butterflies, 63 percent shifted their ranges to the north during the 20th century and only 3 percent expanded south (Parmesan et al. 1999). No similar study has been done in the Northeast. However, field work suggests that southern species such as the giant swallowtail, white M hairstreak, and red-banded hairstreak, and the sachem already appear to be expanding northward into the region (Wagner 2007), and models predict that northern species such as the Atlantis fritillary and arctic skipper will eventually be extirpated.

Insects enter into many ecological relationships with plants and other animals, and changes in these relationships could be of great consequence. Of particular importance would be changes in plant-herbivore, predator-prey, parasite-host, and pollinator-plant relationships. For example, pollinating insects are vital to the reproduction of flowering plants and climate change has been linked to declining pollinator abundance (NAS 2007). The presence and strength of such interactions in the Northeast should be the focus of future research.

HOW ARE NUISANCE SPECIES OF NORTHEAST FORESTS RESPONDING TO CLIMATE CHANGE?

Summary

Pests, pathogens, and invasive plants are already among the leading causes of disturbance to forests of North America and may become more severe as climate changes. Reviews of the known biological responses of six nuisance species predict that five will become more widespread, more abundant, or have more severe impacts in Northeast forests as the climate changes (Table 6). The increase in the severity and expanded range of nuisance species will likely add to the stress that forest ecosystems will face in the next century.

Forest Pests: Observed and Projected Changes

Outbreaks of forest insect pests such as the Asian longhorned beetle, hemlock woolly adelgid, and forest tent caterpillar are of great ecological importance. They can reduce the vitality of trees and damage foliage, resulting in widespread tree mortality. Tree mortality can affect surface water quality and wildlife populations.

Climate change will likely reduce mortality of many pest species and can lead to expansions and outbreaks. Projected reductions in the frequency and intensity of extremely cold temperatures is especially important, as it will increase winter survival and allow many insects to expand their range northward. Already, extreme minimum temperatures have increased by 5.9 °F (3.3 °C) in the southeastern United States between 1960 and 2004, and outbreaks of southern pine beetle have extended northward by about 200 km, matching climate-based predictions (Tran et al. 2007).

Case Study: Hemlock Woolly Adelgid

The case of the hemlock woolly adelgid provides useful insights into the potential response of forest insect pests in the Northeast. The adelgid is an introduced aphid-like insect from Japan that feeds on and kills eastern hemlock. In the past 10 years, the adelgid has infested many hemlock stands in the southern part of the Northeast. If it continues to expand northward, it may produce a range-wide decline in, or possible elimination of, this important tree species.

The northerly spread and ultimate range of the adelgid will likely be controlled by the severity, duration, and timing of minimum winter temperatures. Currently, the adelgid is restricted to

Table 6.—Modeled responses of six nuisance species to climate warming. Impact refers to the severity of impact within the three species range. Adapted from Dukes et al 2009.

	Range	Impact	Confidence
Hemlock woolly adelgid	+	+	high
Tent caterpillar	+ or 0	unknown	medium
Root rot	0	+	medium
Beech bark disease	+	unknown	medium
Oriental bittersweet	+	0	low
Glossy buckthorn	0	0	low

areas where minimum winter temperatures stay above -20 °F (-29 °C). In a study of 36 sites across the Northeast, adelgid mortality was positively correlated with latitude and minimum temperatures recorded per site. Its cold hardiness also depends on time of year; the insects lose their ability to tolerate cold as the winter progresses (Skinner et al. 2003). Thus not only the severity but the timing of cold events is critical. If warming occurs as predicted, milder winters may remove the current limits to the adelgid's range, and increased survival and fecundity may result in larger populations.

Forest Pathogens: Observed and Projected Changes

Forest pathogens such as Dutch elm disease, beech bark disease, and *Armillaria* root rot can have adverse impacts on forest structure and species composition, and alter ecosystem function. Pathogens commonly benefit from increases in temperature and precipitation, and are generally adaptable, suggesting that the impact of forest pathogens may increase in the next few decades.

Forest pathogens may be fungal, viral, or bacterial. The fungi are the best known. Fungal pathogens can survive and remain infective over a wide range of temperatures. However, the conditions that favor epidemic growth for most fungal pathogens are constrained to within a band of a few degrees Celcius. Less is known about viral or bacterial sensitivity to climate in forest systems, but infection and transmission rates seem to vary with temperature and moisture. Higher minimum winter temperatures may favor the winter survival of some; decreased snow cover may increase their exposure to lower temperatures. Increased rainfall will favor many forest pathogens by enhancing spore production and dispersal by rain splash. But the infectivity of some, such as powdery mildew, is decreased by high moisture, and so the increased frequency of summer drought will favor some pathogens and be a challenge to others.

The genetic adaptability of forest pathogens will be important under changing climate conditions. The generation times of pathogens are much shorter than those of their tree hosts, allowing them to respond faster as the climate changes. Theory predicts that an introduced pathogen that has encountered a novel host and is undergoing sustained population growth should exhibit a more rapid evolutionary response to changing environmental conditions than a native pathogen (Brasier 1995). If introduced pathogens can take advantage of changing environmental conditions in this way, it could have major impacts on tree health and survivorship.

Case Study: *Armillaria* Root Rot

Armillaria is a common root and tree butt pathogen. It is currently widespread in eastern deciduous forests. Typically, it is a secondary pathogen, killing only weakened or stressed hosts. As such it contributes to structural diversity, provides habitat for wildlife and microbes, and aids in the recycling of nutrients. Since *Armillaria* is already well established across the Northeast, the effects of climate change on its further dispersal are of minor interest. What is of concern is how it will respond as forest trees are stressed by climate. For example, it is possible that higher annual temperatures, especially during winter months, will allow *Armillaria* to remain active for more of the year. Higher summer temperatures coupled with more frequent and severe drought may allow *Armillaria* to increase its colonization of live hosts, especially where additional stresses like insect defoliation are present. These possibilities suggest that *Armillaria* may both expand its range and become more aggressive as the climate changes (Dukes et al. 2009).

Invasive Plants: Observed and Projected Changes

The introduction of invasive plants such as Norway maple, garlic mustard and oriental bittersweet has been affecting Northeast forests for over a century. While dozens of introduced species occur in forests, the greatest impacts have come from a suite of about 10 fruit-bearing shrubs and vines. Japanese barberry, European honeysuckle, oriental bittersweet, and the two European buckthorns are among the most common. They have greatly modified forest understories in the Northeast, particularly in young or over-browsed or physically disturbed forests. They typically form dense thickets, effectively eliminating tree regeneration and reducing native understory shrub and herb diversity.

Garlic mustard in flower.
Photo by Chris Evans, River to River CWMA, Bugwood.org

These introduced shrubs and vines have multiple impacts. They may alter soil chemistry (Japanese barberry; Ehrenfeld et al. 2001), reduce plant diversity and native butterfly populations (garlic mustard; Stinson et al. 2006), and choke native saplings and trees (oriental bittersweet; Steward et al. 2003).

As with pests and pathogens, future climate change is likely to alter the range and abundance of invasive plants. Several lines of reasoning suggest that as a group, invasive plant species could benefit more from climate change than native plants. Invasive plant species are often better able to tolerate or adjust to new climates than native species. They often have broad environmental tolerance (Qian and Ricklefs 2006), are able to change form and structure in response to changing environmental conditions (Schweitzer and Larson 1999), and are capable of rapid evolutionary change (Maron et al. 2004). All of these properties could allow invasives to maintain or even increase their fitness relative to other species in a changing climate. But, as with many aspects of climate change biology, the adaptive capacity of the invasives is largely untested, and the hypothesized fitness gains have not actually been observed.

Japanese barberry infestation.
Photo by Leslie J. Mehrhoff, University of Connecticut, Bugwood.org

Case Study: Oriental bittersweet

Oriental bittersweet is a twining vine from southeastern Asia that is abundant in southern New England, but rare in northern New England and Canada. Where it occurs, it is a severe pest that has profound impacts on forests, damaging trees by girdling their trunks, breaking tree branches, and shading out young saplings. It can form a secondary canopy over the tops of trees and suppress flowering and reproduction. This may prevent forest succession – posing an additional challenge to forests adapting to climate change.

The available data suggest that bittersweet is likely to benefit from the warming and increased precipitation that are predicted for the Northeast. The species has broad habitat tolerances and is able to grow well in a wide range of conditions, all of which will enhance its success in a changing climate (Clement et al. 1991, Ellsworth et al. 2004, Patterson 1974). Results from computer modeling studies predict that oriental bittersweet has the potential to spread throughout New England, and that the likelihood of its invading parts of northern New England is high (Dukes et al. 2009). This is corroborated by research which shows that it does well in frequently harvested mesic forests with significant or frequent wind disturbance, a frequently scarified forest floor, and gaps in the forest canopy (McNab and Loftis 2002). Such forests are common in northern New England.

Oriental bittersweet.
Photo by Linda Haugen, USDA Forest Service, Bugwood.org

Oriental bittersweet infestation.
Photo by Chris Evans, River to River CWMA
Bugwood.org

CONCLUSIONS AND IMPLICATIONS FOR POLICY: MITIGATING AND ADAPTING TO CLIMATE CHANGE IN NORTHEAST FORESTS

The NE Forests 2100 research initiative brought together forest scientists from across the northeastern United States and eastern Canada. The knowledge assembled through their efforts demonstrates that the climate of the Northeast has changed and will likely continue to change. This report shows the diverse and profound impacts climate changes are already having on the region's forests and forecasts the effects of future change. The projections of future impacts suggest that forests will have trouble keeping pace with the accelerating rate of climate change and the associated stresses that climate change generates.

Given that changes are likely coming, how should forest managers, policy makers, nongovernmental organizations, scientists, and concerned citizens respond? This last section draws on the work of NE Forest 2100 research and others to outline some principals for decision makers.

Policy and management options for addressing climate change and its impacts fall into two categories: mitigation and adaptation. Mitigation refers to "an anthropogenic intervention to reduce the sources or enhance the sinks of greenhouse gases" (IPCC 2007). Adaptation is "an adjustment in natural or human systems in response to actual or expected climatic stimuli or their effects, which moderates harm or exploits beneficial opportunities" (IPCC 2007). Three central ideas each for mitigation and adaptation emerge from recent forest science and policy reports and are outlined below.

Mitigation I: Prevent Forest Loss

Several recent reports in the Northeast and wider United States suggest that the first and best action to help mitigate climate change is to conserve existing forest land. A recent report by the Hubbard Brook Research Foundation presented carbon budgets for eight counties in New England. Their results show that heavily forested areas sequester more carbon than they emit, whereas developed lands are net sources of CO_2 to the atmosphere (Fahey et al. 2011). Suburban sprawl, which has its high emissions from transportation, was identified as the land use with the largest carbon footprint, even greater than dense urban areas (Fahey et al. 2011). The Hubbard Brook Research Foundation study also showed that forests, despite their variability, provided far greater benefits to climate stabilization than the alternative of land development. The cost of conserving forest land to secure carbon storage in perpetuity may be viewed as expensive. However, when additional benefits of forest protection—such as clean drinking water, flood protection, wildlife habitat, and recreational opportunities—are considered, the investment can be quite attractive.

Another recent study in the region (Foster et al. 2010) looked at the history of forest cover in New England and offered a preservation plan. The report, which was produced at the Harvard Forest, MA, a research forest associated with Harvard University, says that the regrowth of New England's forests following a century of land clearing by early settlers is both an opportunity and a great challenge. New England is currently 80 percent forested, and its forests are globally significant for carbon storage. In every New England state, development is decreasing forest cover and reducing carbon storage. The authors call for retaining at least 70 percent of the region as forest land. Because 80 percent of the forests of New England are privately owned, achieving the vision will require conserving private woodlands for multiple uses through a combination of conservation easements, strategic acquisition, and enhanced economic incentives for private stewardship. National studies have echoed the call to conserve forests. The Ecological Society of America lists avoiding deforestation as a high priority and calls for conserving forests (Ryan et al. 2010). Avoiding deforestation is a low-risk way of retaining forest carbon and this option is a consideraton for inclusion in state and regional conservation initiatives and climate change mitigation programs.

Mitigation II: Enhance Carbon Storage in Managed Forests

A second strategy for using forests to mitigate climate change is the enhancement of carbon sequestration by forest management (Canadell and Raupach 2008, Ryan et al. 2010). The amount of carbon stored in Northeast forests is roughly 7 billion metric tons. This could be increased by lengthening the harvest interval, reducing the amount of wood removed in each harvest, and increasing the rate of forest growth through intensive silviculture (Ryan et al. 2010).

Just how much additional carbon can be stored by improved management is uncertain. Some scientists believe that the region's forests have peaked in their carbon storage potential. Others suggest that forests continue to store carbon as they age, especially in soils and coarse woody debris (Foster et al. 2010, Luyssaert et al. 2008). Still others argue that managed forests can reduce net greenhouse gas emission, if they are sustainably managed and if the products from them are used to replace carbon-intensive building materials such as concrete (Canadell and Raupach 2008, Ryan et al. 2010).

Mitigation III: Replace Fossil Fuels with "Smart" Biomass

Sustainably managed forests can supply woody biomass for energy production. Most scientists agree that the displacement of fossil fuel by wood from existing harvests is likely to result in a net reduction in greenhouse gas emissions, provided that the wood is harvested sustainably and used in efficient applications such as community-scale combined heat and power biomass energy systems. Wood biomass projects can provide additional income to forest landowners and may prevent or defer the conversion of forests to other land uses and thereby prevent the emissions associated with forest conversion. However, the carbon benefits from biomass energy production are not guaranteed, and will only be secured if the forest management is sustainable and energy generation is efficient.

Adaptation I: Increase Protected Areas

Adaptation strategies for Northeast forests would benefit from allowing the movement of native plant and animal species in response to climate change, helping maintain ecosystem function, and conserving opportunities for adaptation of species to modified climate conditions (Heinz Center 2008). As with mitigation strategies, the protection of forested landscapes, and in particular forested natural areas, tops the list of priorities. The first challenge will be to find the resources to do this. The second will be to create a network of protected areas that will be effective at conserving dynamic patterns of biodiversity at a variety of scales (Heinz Center 2008, Lovejoy 2005). Building a portfolio of protected forest areas that conserve more than one example of all important forest types in the region is also recommended (Julius and West 2007).

Adaptation II: Conserve Stepping Stones, Corridors and Refuges

Climate adaptation will benefit from a landscape-scale conservation perspective that explicitly considers how the individual parcels of conserved land link to one another across large regions. Ecologists believe that "stepping stones", which are corridors and habitat islands that link larger reserves, will aid the movement of wildlife and plants under changing climate regimes (Heinz Center 2008).

The management of corridors in large landscapes requires focusing on the areas that are likely to be important dispersal pathways (Carroll 2005). Scientists have long recognized that some environments are more buffered against climate change than others (Millar et al. 2007). During periods of historical change, these places act as refuges. Refuges provide conditions where plant and animal populations are able to persist due in part to the ability of these local sites to escape the extremes of regional climate change impacts (Millar et al. 2007). Conserving known refuges would allow populations to persist and eventually colonize new suitable areas, if conditions permit.

Given that most forest land in the Northeast is privately owned, the effectiveness of these landscape-scale strategies depends directly on the cooperation and engagement of the hundreds of thousands of family forest owners across the region.

Adaptation III: Reduce Other Stresses on Forests

The IPCC identified the importance of reducing other environmental threats as a way of adapting to the complex issue of climate change. These other threats include habitat fragmentation or loss, pollution (including acid deposition and nitrogen enrichment), over-exploitation of natural resources, and the introduction of alien species (Fischlin et al. 2007). Reducing these other stresses may increase the ability of forest (and other) ecosystems to tolerate climate change (Julius and West 2007).

ACKNOWLEDGMENTS

Portions of this report are built on work originally developed by a coalition of U.S. and Canadian scientists as part of the NE Forests 2100 initiative. The results of that effort were published in a series of papers in the Canadian Journal of Forest Research (2009). The authors thank Steven McNulty, MaryBeth Adams, Richard Birdsey, Jamie Shandley, and Jerry Jenkins for their thoughtful comments on the manuscript. The activities of the NE Forests 2100 initiative have been supported by grants from the Northeastern States Research Cooperative (NSRC) and by organizing efforts of the Northeastern Ecosystem Research Cooperative (NERC). The development of this report was generously supported by U.S. Forest Service, Northern Research Station and the New York State Energy Research and Development Authority (NYSERDA).

Waterfall in southern New England. Photo by Brett J. Butler, U.S. Forest Service.

LITERATURE CITED

Arft, A.M.; Walker, M.D.; Gurevitch, J.; Alatalo, J.M.; Bret-Harte, M.S.; Dale, M.; Dimer, M.; Guggerli, F.; Henry, G.H.T.; Jones, M.H.; Hollister, R.D.; Jonsdottir, I.S.; Laine, K.; Levesque, E.; Marion, G.M.; Molau, U.; Molgaard, P.; Nordenhall, U.; Raszhivin, V.; Robinson, C.H.; Starr, G.; Stenstrom, A.; Stenstrom, M.; Totland, O.; Turner, P.L.; Walker, L.J.; Webber, P.J.; Welker, J.M.; Wookey, P.A. 1999. **Responses of tundra plants to experimental warming: meta-analysis of the international tundra experiment.** Ecological Monographs. 69: 491-511.

Balch, R.E. 1944. **The dieback of birch in the Maritime Region**. Contribution No. 3. Fredericton, NB: Dominion of Canada, Department of Agriculture, Division of Entomology. [mimeo].

Beckage, B.; Osborne, B.; Gavin, D.G.; Pucko, C.; Siccama, T.; Perkins, T. 2008. **A rapid upward shift of a forest ecotone during 40 years of warming in the Green Mountains of Vermont.** Proceedings of the National Academy of Sciences. 105(11): 4197-4202.

Bertrand, A.; Robitaille, G.; Nadeau, P.; Boutin, R. 1994. **Effects of soil freezing and drought stress on abscisic acid content of sugar maple sap and leaves.** Tree Physiology. 14(4): 413-425.

Bourque, C.P.A.; Cox, R.M.; Allen, D.J.; Arp, P.A.; Meng, F.-R. 2005. **Spatial extent of winter thaw events in eastern North America: historical weather records in relation to yellow birch decline.** Climate Change Biology. 11: 1477-1492.

Braathe, P. 1995. **Birch dieback caused by prolonged early spring thaws and subsequent frost.** Norwegian Journal of Agricultural Science. 20 (Suppl.): 50-109.

Brasier, C.M. 1995. **Episodic selection as a force in fungal microevolution, with special reference to clonal speciation and hybrid introgression.** Canadian Journal of Botany. 73: S1213-S1221.

Brooks, R.T. 2004. **Weather-related effects on woodland vernal pool hydrology and hydroperiod.** Wetlands. 24: 104-114.

Burakowski, E.A.; Wake, C.P.; Braswell, B.; Brown, D.P. 2008. **Trends in wintertime climate in the northeastern United States 1965-2005.** Journal Geophysical Research. 113: D20114. doi: 20110.21029/2208JD009870.

Burton, A.J.; Pregitzer, K.S.; Zogg, G.P.; Zak, D.R. 1998. **Drought reduces rot respiration in sugar maple forests.** Ecological Applications. 8: 771-778.

Campbell, J.L.; Rustad, L.E.; Boyer, E.W.; Christopher, S.F.; Driscoll, C.T.; Fernandez, I.J.; Groffman, P.M.; Houle, D.; Kiekbusch, J.; Magill, A.H.; Mitchell, M.J.; Ollinger,

S.V. 2009. **Consequences of climate change for biogeochemical cycling in forests of northeastern North America.** Canadian Journal of Forest Research. 39: 264-284.

Canadell, J.G.; Raupach, M.R. 2008. **Managing forests for climate change mitigation.** Science. 320: 1456-1457.

Carroll, C. 2005. **Carnivore restoration in the northeastern U. S. and southeastern Canada: A regional-scale analysis of habitat and population viability for wolf, lynx, and marten; Report 2: lynx and marten viability analysis.** Richmond, VA: The Wildlands Project. 46 p.

Chapman, S.J.; Thurlow, M. 1996. **The influence of climate on CO2 and CH4 emissions from organic soils.** Agriculture and Forest Meteorology. 79: 205-217.

Christensen, T.R.; Ekberg, A.; Ström, L.; Mastepanov, M.; Panikov, N.; Öquist, M.; Svensson, B.H.; Nykänen, H.; Martikainen, P.J.; Oskarsson, H. 2003. **Factors controlling large scale variations in methane emission from wetlands.** Geophysical Research Letters. 30: 1414-1419.

Clement, C.; Warren, R.; Dreyer, G.; Barnes, P. 1991. **Photosynthesis water relations and fecundity in the woody vines American and Oriental bittersweet (*Celastrus scandens* and *C. orbiculatus*).** American Journal of Botany. 78(Suppl. 6): 134-134. [Abstract].

Decker, K.L. M.; Wang, D.; Waite, C.; Scherbatskoy, T. 2003. **Snow removal and ambient air temperature effects on forest soil temperatures in northern Vermont.** Soil Science Society of America Journal. 67: 1234-1242.

DeGaetano, A.T.; Allen, R.J. 2002. **Trends in twentieth-century temperature extremes across the United States.** Journal of Climatology. 15: 3188-3205.

Dukes, J.S.; Pontius, J.; Orwig, D.; Garnas, J.R.; Rodgers, V.L.; Brazee, N; Cooke, B.; Theoharides, K.A.; Stange, E.E.; Harrington, R.; Ehrenfeld, J.; Gurevitch, J.; Lerdau, M.; Stinson, K.; Wick, R.; Ayres, M. 2009. **Responses of insect pests, pathogens, and invasive plant species to climate change in the forests of northeastern North America: What can we predict?** Canadian Journal of Forest Research. 39: 231-248.

Ehrenfeld, J.G.; Kourtev, P.; Huang, W.Z. 2001. **Changes in soil functions following invasions of exotic understory plants in deciduous forests.** Ecological Applications. 11: 1287-1300.

Ellison, A.M.; Bank, M.S.; Clinton, B.D.; Colburn, E.A.; Elliott, K.; Ford, C.R.; Foster, D.R.; Knowpp, J.D.; Lovett, G.M.; Mohan, J.E.; Orwig, D.A.; Rodenhouse, N.L.; Sobczak, W.V.; Stinson, K.A.; Stone, J.K.; Swank, C.M.; Thompson, J.; von Holbe, D.; Webster, J.R. 2005. **Loss of foundation species: consequences for the structure and dynamics of forested ecosystems.** Frontiers in Ecology and the Environment. 3(9): 479-486.

Ellsworth, J.W.; Harrington, R.A.; Fownes, J.H. 2004. **Seedling emergence, growth, and allocation of oriental bittersweet: effects of seed input, seed bank, and forest floor litter.** Forest Ecology and Management. 190(2-3): 255-264.

Fahey, T.J.; Carranti, F.; Driscoll, C.; Foster, D.; Gwyther, P.; Hall, B.; Hamburg, S.; Jenkins, J.C.; Jenkins, J.; Neill, C.; Ollinger, S.; Peery, B.; Quigley, E.; Raciti, S.; Sherman, R.; Thomas, R.; Vadeboncoeur, M.; Weinstein, D.; Wilson, G.; Woodbury, P.; Yandik, W. 2011. **Carbon and communities: Linking carbon science with public policy and resource management in the northeastern United States.** Science Links Publication 1(4). North Woodstock, NH: Hubbard Brook Research Foundation.

Fernandez, I.J.; Rustad, L.E.; Norton, S.A.; Kahl, J.S.; Cosby, B.J. 2003. **Experimental acidification causes soil base-cation depletion at the Bear Brook Watershed in Maine.** Soil Science Society of America Journal. 67: 1909-1919.

Fischlin, A.; Midgley, G.F.; Price, J.T.; Leemans, R.; Gopal, B.; Turley, C.; Rounsevell, M.D.A.; Dube, O.P.; Tarazona, J.; Velichko, A.A. 2007. **Ecosystems, their properties, goods, and services.** In: Parry, M.L.; Canziani, O.F.; Palutikof, J. P.; van der Linden, P.J.; Hanson, C.E., eds. Climate change 2007: Impacts, adaptation and vulnerability. Contribution of Working Group II to the fourth assessment report of the Intergovernmental Panel on Climate Change. Cambridge: Cambridge University Press: 211-272.

Fitzhugh, R.D.; Driscoll, C.T.; Groffman, P.M.; Tierney, G.L.; Fahey, T.J.; Hardy, J.P. 2001. **Effects of soil freezing disturbance on soil solution nitrogen, phosphorus, and carbon chemistry in a northern hardwood ecosystem.** Biogeochemistry. 56: 215-238.

Fitzhugh, R.; Driscoll, C.; Groffman, P.; Tierney, G.; Fahey, T.; Hardy, J. 2003. **Soil freezing and the acid-base chemistry of soil solutions in a northern hardwood forest.** Soil Science Society of America Journal. 67(6): 1897-1908.

Foster, D.R.; Donahue, B.M.; Kittredge, D.B.; Lambert, K.F.; Hunter, M.L.; Hall, B.R.; Irland, L.C.; Lilieholm, R.J.; Orwig, D.A.; D'Amato, A.W.; Colburn, E.A.; Thompson, J.R.; Levitt, J.N.; Ellison, A.M.; Keeton, W.S.; Aber, J.D.; Cogbill, C.V.; Driscoll, C.T.; Fahey, T.J.; Hart, C.M. 2010. **Wildlands and woodlands: a vision for the New England landscape.** Petersham, MA: Harvard Forest, Harvard University. 36 p.

Friedland, A.J.; Gregory, R.A.; Karenlampi, L.A.; Johnson, A.H. 1984. **Winter damage to foliage as a factor in red spruce decline.** Canadian Journal of Forest Research. 14: 963-965.

Friedlingstein, P.; Houghton, R.A.; Marland, G.; Hackler, J.; Boden, T.A.; Conway, T.J.; Canadell, J.G.; Raupach, M.R.; Ciais, P.; Le Quéré, C. 2010. **Update on CO2 emissions.** Nature Geoscience. 3: 811-812.

Frumhoff, P.C.; McCarthy, J.J.; Melillo, J.M.; Moser, S.C.; Wuebbles, D.J. 2007. **Confronting climate change in the U.S. northeast: Science, impacts, and solutions.**

Synthesis report of the Northeast Climate Impacts Assessment (NECIA). Cambridge, MA: Union of Concerned Scientists.146 p.

Gibbs, J.P.; Breisch, A.R. 2001. **Climate warming and calling phenology of frogs near Ithaca, New York, 1900-1999.** Conservation Biology. 15: 1175-1178.

Groffman, P.M.; Hardy, J.P.; Driscoll, C.T.; Fahey, T.J. 2006. **Snow depth, soil freezing, and trace gas fluxes in a northern hardwood forest.** Global Change Biology. 12: 1748-1760.

Hartley, S.; Dingman, S.L. 1993. **Effects of climatic variability on winter-spring runoff in New England river basins.** Physiology Geography. 14: 379-393.

Hayhoe, K.; Wake, C.P.; Huntington, T.G.; Luo, L.; Schwartz, M.D.; Sheffield, J.; Wood, E.; Anderson, B.; Bradbury, J.; DeGaetano, A.; Troy, T.J.; Wolf, D. 2007. **Past and future changes in climate and hydrological indicators in US Northeast.** Climate Dynamics. 28: 381-407.

Heinz Center. 2008. **Strategies for managing the effects of climate change on wildlife and ecosystems.** Washington, DC: The Heinz Center. 43 p.

Hodgkins, G.A.; Dudley, R.W. 2006. **Changes in late-winter snowpack depth, water equivalent, and density in Maine, 1926-2004.** Hydrological Processes. 20: 741-751.

Hodgkins, G.A.; Dudley, R.W. 2005. **Changes in the magnitude of annual and monthly streamflows in New England, 1902-2002.** Scientific Investigations Report. 2005-5235. Reston, VA: U.S. Department of Interior, Geological Survey.

Hodgkins, G.A.; Dudley, R.W.; Huntington, T.G. 2005. **Changes in the number and timing of ice-affected flow days on New England rivers, 1930-2000.** Climate Change. 71: 319-340.

Hodgkins, G.A.; Dudley, R.W.; Huntington, T.G. 2003. **Changes in the timing of high river flows in New England over the 20th century.** Journal of Hydrology. 278: 244-252.

Hodgkins, G.A.; James, I.C.; Huntington, T.G. 2002. **Historical changes in lake ice-out dats as indicators of climate change in New England.** International Journal of Climatology. 22: 1819-1827.

Humphries, M.M.; Umbanhowar, J.; McCann, K.S. 2004. **Bioenergetic prediction of climate change impacts on northern mammals.** Integrative and Comparative Biology. 44: 152-162.

Huntington, T.G. 2003. **Climate warming could reduce runoff significantly in New England, 2003.** Agricultural and Forest Meteorology. 117: 193-201.

Huntington, T.G. 2006. **Evidence for intensification of the global water cycle: review and synthesis.** Journal of Hydrology. 319: 83-95.

Huntington, T.G.; Hodgkins, G.A.; Dudley, R.W. 2003. **Historical trend in river ice thickness and coherence in hydroclimatological trends in Maine.** Climate Change. 61: 217-236.

Huntington, T.G.; Hodgkins, G.A.; Keim, B.D.; Dudley, R.W. 2004. **Changes in the proportion of precipitation occurring as snow in New England (1949-2000).** Journal of Climate. 17: 2626-2636.

Huntington, T.G.; Richardson, A.D.; McGuire, K.J.; Hayhoe, K. 2009. **Climate and hydrological changes in the northeastern United States: recent trends and implications for forested and aquatic ecosystems.** Canadian Journal of Forest Research. 39: 199-212.

International Snowmobile Manufacturers Association. 2006. **International snowmobile industry facts and figures.** Available from: http://www.snowmobile.org/pr_snowfacts.asp.

Intergovernmental Panel on Climate Change (IPCC). 2007. **Climate Change 2007: The physical science basis.** Contribution of Working Group I to the Fourth Assessment Report of the Intergovernmental Panel on Climate Change. [Solomon, S.; Qin, D.; Manning, M.; Chen, Z.; Marquis, M.; Avery, K.B.; Tignor, M.; Miller, H.L., eds.] Cambridge, UK: Cambridge University Press. 996 p.

Iverson, L.R.; Prasad, A.; Matthews, S. 2008. **Modeling potential climate change impacts on the trees of the northeastern United States.** Mitigation and Adaptation Strategies for Global Change. 13: 517-540.

Johnson, A.H. 1992. **The role of abiotic stresses in the decline of red spruce in high elevation forests of the eastern United States.** Annual Review of Phytopathology. 30: 349-369.

Julius, S.H.; West, J.M. 2007. **Preliminary review of adaptation options for climate-sensitive ecosystems and resources, synthesis and assessment product 4.4.** U.S. Climate Change Science Program, draft for public comment—August 2007. Available at: http://www.climatescience.gov/Library/sap/sap4-4/default.php

Keim, B.D.; Fischer, M.R.; Wilson, A.M. 2005. **Are there spurious precipitation trends in the United States Climate Division database?** Geophysical Research Letters. 32: L04702. doi:10.1029/204GL021985.

Lambert, J.D.; McFarland, K.P.; Rimmer, C.C.; Faccio, S.D.; Atwood, J.L. 2005. **A practical model of Bicknell's Thrush distribution in the northeastern United States.** Wilson Bulletin. 117: 1-11.

Likens, G.E.; Bormann, F.H. 1995. **Biogeochemistry of a forested ecosystem.** 2nd edition. New York, NY: Springer-Verlag. 171 p.

Lovejoy, T.E. 2005. **Conservation with a changing climate.** In: Lovejoy, T. E.; Hannah, L., eds. Climate change and biodiversity. New Haven and London: Yale University Press: 325-328.

Luyssaert, S.; Schulze, E.D.; Börner, A.; Knohl, A.; Hessenmöller, D.; Law, B.E.; Ciais, P.; Grace, J. 2008. **Old-growth forests as global carbon sinks.** Nature. 455: 213-215.

Maron, J.L.; Vila, M.; Bommarco, R.; Elmendorf, S.; Beardsley, P. 2004. **Rapid evolution of an invasive plant.** Ecological Monographs. 74(2): 261-280.

Matthews, S.N.; O'Connor, R.J.; Iverson, L.R.; Prasad, A.M. 2004. **Atlas of climate change effects in 150 bird species of the eastern United States.** Gen. Tech. Rep. NE-318. Newtown Square, PA: U.S. Department of Agriculture, Forest Service, Northeastern Research Station. 340 p.

McCormick, S.D.; Hansen, L.P.; Quinn, T.P.; Saunders, R.L. 1998. **Movement, migration, and smolting of Atlantic salmon (*Salmo salar*).** Canadian Journal of Fisheries and Aquatic Sciences. 55: 77-92.

McHale, P.J.; Mitchell, M.J.; Bowles, F.P. 1998. **Soil warming in a northern hardwood forest: trace gas fluxes and leaf litter decomposition.** Canadian Journal of Forest Research. 28: 1365-1372.

McLaughlin, S.B.; Nosal, M.; Wullschleger, S.D.; Sun, D. 2007. **Interactive effects of ozone and climate on tree growth and water use in a southern Appalachian forest in the USA.** New Phytologist. 174: 109-124.

McNab, W.H.; Loftis, D.L. 2002. **Probability of occurrence and habitat features for oriental bittersweet in an oak forest in the southern Appalachian mountains, USA.** Forest Ecology and Management. 155: 45-54.

Michaels, G.; O'Neal, K.; Humphrey, J.; Bell, K.; Camacho, R.; Funk, R. 1995. **Ecological impacts from climate change: An economic analysis of freshwater fishing.** EPA 220-R-95-004. Washington, DC: U.S. Environmental Protection Agency, Office of Policy, Planning, and Evaluation. 230 p.

Millar, C.I.; Stephenson, N.L.; Stephens, S.L. 2007. **Climate change and Forests of the Future: Managing in the Face of Uncertainty.** Ecological Applications. 17(8): 2145-2151.

Millers, I.; Shriner, D.S.; Rizzo, D. 1989. **History of hardwood decline in the eastern United States.** Gen. Tech. Rep. NE-216. Broomall, PA: U.S. Department of Agriculture, Forest Service, Northeastern Forest Experiment Station. 75 p.

Mohan, J.E.; Cox, R.M.; Iverson, L.R. 2009. **Composition and carbon dynamics of forests in northeastern North America in a future, warmer world.** Canadian Journal of Forest Research. 39: 213-230.

National Academy of Science. 2007. **Status of pollinators in North America.** Washington, DC: National Research Council of the National Academy of Science, Committee on the Status of Pollinators in North America. 312 p.

Nakicenovic, N.; Alcamo, J.; Davis, G.; de Vries, B.; Fenhann, J.; Gaffin, S.; Gregory, K.; Grubler, A.; Jung, T.Y.; Kram, T.; La Rovere, E.L.; Michaelis, L.; Mori, S.; Morita, T.; Pepper, W.; Pitcher, H.; Price, L.; Riahi, K.; Roehrl, A.; Rogner, H.-H.; Sankovski, A.; Schlesinger, M.; Shukla, P.; Smith, S.; Swart, R.; van Rooijen, S.; Victor, N.; Dadi, Z. 2000. **Special report on emissions scenarios.** An Intergovernmental Panel on Climate Change Report. Cambridge, UK: Cambridge University Press. 599 p.

New England Regional Assessment Group (NERA). 2001. **Preparing for a changing climate: The potential consequences of climate variability and change.** New England Regional Overview, U.S. Global Change Research Program. Durham, NH: University of New Hampshire. 96 p.

Ollinger, S.V.; Goodale, C.L.; Hayhoe, K.; Jenkins, J.P. 2008. **Potential effects of climate change and rising CO_2 on ecosystem processes in northeastern U.S. forests.** Mitigation and Adaptation Strategies for Global Change. 13: 467-485.

Palatova, E. 2002. **Effects of increased nitrogen deposition and drought stress on the development of Scots pine (*Pinus sylvestris*). II: Root system response.** Journal of Forest Science. 48: 237-247.

Parmesan, C.; Ryrholm, N.; Stefanescu, C.; Hill, J.K.; Thomas, C.D.; Descimon, H.; Huntley, B.; Kaila, L.; Kullberg, J.; Tammaru, T.; Tennent, W.J.; Thomas, J.A.; Warren, M. 1999. **Poleward shifts in geographical ranges f butterfly species associated with regional warming.** Nature. 399: 579-583.

Patterson, D.T. 1974. **The ecology of oriental bittersweet, *Celastrus orbiculatus*, a weedy introduced ornamental vine.** Durham, NC: Duke University. Ph.D. dissertation.

Peterjohn, W.T.; Melillo, J.M.; Steudler, P.A.; Newkirk, K.M.; Bowles, F.P.; Aber, J.D. 1994. **Responses of trace gas fluxes and N availability to experimentally elevated soil temperatures.** Ecological Applications. 4: 617-625.

Poland, T.M.; McCullough, D.G. 2006. **Emerald ash borer: invasion of the urban forest and the threat to North America's ash resource.** Journal of Forestry. 104: 118-124.

Prowse, T.D.; Beltaos, S. 2002. **Climatic control of river-ice hydrology: a review.** Hydrological Processes. 16: 805-822.

Qian, H.; Rickelfs, R.E. 2006. **The role of exotic species in homogenizing the North American flora.** Ecological Letters. 9: 1293-1298.

Rahbek, C.; Gotelli, N.J.; Colwell, R.K.; Entsminger, G.L.; Fernando, T.; Rangel, L.V.B.; Graves, G.R. 2007. **Predicting continental scale patterns in bird species richness with spatially explicit models.** Proceedings of the Research Society of London Biological Sciences. 274: 165-174.

Renecker. L.A.; Hudson, R.J. 1986. **Seasonal energy expenditures and thermoregulatory responses of moose.** Canadian Journal of Zoology. 64: 322-327.

Richardson, A.D.; Bailey, A.S.; Denny E.G.; Martin C.W.; O'Keefe, J. 2006. **Phenology of a northern hardwood forest canopy.** Global Change Biology. 12: 1174-1188.

Rodenhouse, N.L.; Matthews, S.N.; McFarland, K.P.; Lambert, J.D.; Iverson, L.R.; Prasad, A.; Sillett, T.S.; Holmes, R.T. 2008. **Potential effects of climate change on birds of the Northeast.** Mitigation and Adaption Strategies for Global Change. 13: 517-540.

Rodenhouse, N.L.; Christenson, L.M.; Parry, D.; Green, L.E. 2009. **Climate change effects on native fauna of northeastern forests.** Canadian Journal of Forest Research. 39: 249-263.

Rustad, L.E.; Fernandez, I.J. 1998. **Soil warming: consequences for foliar litter decay in a spruce-fir forest in Maine, USA.** Soil Science Society of America Journal. 62: 1072-1080.

Rustad, L.E.; Campbell, J.L.; Marion, G.M.; Norby, R.J.; Mitchell, M.J.; Hartley, A.E.; Cornelissen, J.H.C.; Gurevitch, J. GCTE-NEWS. 2001. **A meta-analysis of the response of soil respiration, net nitrogen mineralization, and aboveground plant growth to experimental warming.** Oecologia. 126: 543-562.

Rustad, L.E.; Fernandez, I.J.; Arnold, S. 1996. **Experimental soil warming effects on C and N dynamics in a low elevation spruce-fir forest soil.** In: Hom, J.; Birdsey, R.; O'Brien, K., eds. Proceedings of the 1995 meeting of the Northern Global Climate Change Program; 1995 March 14-16; Pittsburgh, PA. Gen. Tech. Rep. NE-214. Radnor, PA; U.S. Department of Agriculture, Forest Service, Northeastern Research Station: 132-139.

Ryan, M.G.; Harmon, M.E.; Birdsey, R.A.; Giardina, C.P.; Heath, L.S.; Houghton, R.A.; Jackson, R.B.; McKinley, D.C.; Morrison, J.F.; Murray, B.C.; Pataki, D.E.; Skog, K.E. 2010. **A synthesis of the science on forests and carbon for U.S. forests.** Report No. 13. Issues in Ecology. 13: 1-16.

Schaberg, P.G.; DeHayes, D.H. 2000. **Physiology and environmental cause of freezing injury in red spruce.** In: Mickler, R.A.; Birsdey, R.A.; Hom, J., eds. Responses of northern U.S. forests to environmental change. Ecological Studies 139. New York, NY: Springer-Verlag: 181-227.

Schweitzer, J.A.; Larson, K.C. 1999. **Grater morphological plasticity of exotic honeysuckle species may make them better invaders than natives species.** Journal of the Torrey Botanical Society. 126(1): 5-23.

Skimmer, M.; Parker, B.L.; Gouli, S.; Ashikaga, T. 2003. **Regional responses of hemlock woolly adelgid (Homoptera: Adelgidae) to low temperatures.** Environmental Entomology. 32(3): 523-528.

Spierre, S.G.; Wake, C. 2010. **Trends in extreme precipitation events for the northeastern United States, 1948-2007.** Durham, NH: University of New Hampshire, Carbon Solutions New England.

Steward, A.M.; Clemants, S.E.; Moore, G. 2003. **The concurrent decline of the native *Celastrus scandens* and spread of the non-native *Celastrus orbiculatus* in the New York City metropolitan area.** Journal of the Torrey Botanical Society. 130: 143-146.

Stinson, K.A.; Campbell, S.A.; Powell, J.R.; Wolfe, B.E.; Callaway, R.M.; Thelen, G.C.; Hallett, S.G.; Prati, D.; Klironomos, J.N. 2006. **Invasive plant suppresses the growth of native tree seedlings by disrupting belowground mutualisms.** PLoS Biology. 4: 727-731.

Tierney, G.L.; Fahey, T.J.; Groffman, P.; Hardy, J.P.; Fitzhugh, R.D.; Driscoll, C.T. 2001. **Soil freezing alters fine root dynamics in a northern hardwood forest.** Biogeochemistry. 56: 175-190.

Tran, J.K.; Ylioja, T.; Billings, R.F.; Regniere, J.; Ayres, M.P. 2007. **Impact of minimum winter temperatures on the population dynamics of *Dendroctonus frontalis* (Coleoptera: Scolytinae).** Ecological Applications. 17: 882-899.

Trombulak, S.C.; Wolfson, R. 2004. **Twentieth-century climate change in New England and New York, USA.** Journal of Geophysical Research. 31: L19202. doi: 19210.11029/12004GL020574.

Vallee, S.; Payette, S. 2004. **Contrasted growth of black spruce (*Picea mariana*) forest trees in treeline associated with climate change of the last 400 years.** Arctic Antarctica Alpine Research. 36(4): 400-406.

Wagner, D.L. 2007. **Butterfly conservation.** In: O'Donnell, J.E.; Gall, L.F.; Wagner, D.L., eds. Connecticut butterfly atlas. Hartford, CT: Connecticut Department of Environmental Protection: 287-307.

Waite, T.A.; Strickland, D. 2006. **Climate change and the demographic demise of a hoarding bird living on the edge.** Proceedings Research Society of London Biological Sciences. 273: 2809-2813.

Woodall, C.W.; Oswalt, C.M.; Westfall, J.A.; Perry, C.H.; Nelson, M.D.; Finley, A.O. 2009. **An indicator of tree migration in forests of the eastern United States.** Forest Ecology and Management. 257: 1434-1444.

Rustad, Lindsey; Campbell, John; Dukes, Jeffrey S.; Huntington, Thomas; Fallon Lambert, Kathy; Mohan, Jacqueline; Rodenhouse, Nicholas. 2012. **Changing climate, changing forests: The impacts of climate change on forests of the northeastern United States and eastern Canada.** Gen. Tech. Rep. NRS-99. Newtown Square, PA: U.S. Department of Agriculture, Forest Service, Northern Research Station. 48 p.

Decades of study on climatic change and its direct and indirect effects on forest ecosystems provide important insights for forest science, management, and policy. A synthesis of recent research from the northeastern United States and eastern Canada shows that the climate of the region has become warmer and wetter over the past 100 years and that there are more extreme precipitation events. Greater change is projected in the future. The amount of projected future change depends on the emissions scenarios used. Tree species composition of northeast forests has shifted slowly in response to climate for thousands of years. However, current human-accelerated climate change is much more rapid and it is unclear how forests will respond to large changes in suitable habitat. Projections indicate significant declines in suitable habitat for spruce-fir forests and expansion of suitable habitat for oak-dominated forests. Productivity gains that might result from extended growing seasons and carbon dioxide and nitrogen fertilization may be offset by productivity losses associated with the disruption of species assemblages and concurrent stresses associated with potential increases in atmospheric deposition of pollutants, forest fragmentation, and nuisance species. Investigations of links to water and nutrient cycling suggest that changes in evapotranspiration, soil respiration, and mineralization rates could result in significant alterations of key ecosystem processes. Climate change affects the distribution and abundance of many wildlife species in the region through changes in habitat, food availability, thermal tolerances, species interactions such as competition, and susceptibility to parasites and disease. Birds are the most studied northeastern taxa. Twenty-seven of the 38 bird species for which we have adequate long-term records have expanded their ranges predominantly in a northward direction. There is some evidence to suggest that novel species, including pests and pathogens, may be more adept at adjusting to changing climatic conditions, enhancing their competitive ability relative to native species. With the accumulating evidence of climate change and its potential effects, forest stewardship efforts would benefit from integrating climate mitigation and adaptation options in conservation and management plans.

KEY WORDS: temperate forests, biogeochemistry, carbon cycle, water cycle, climate models, invasive species, wildlife, climate adaptation, climate mitigation

www.ingramcontent.com/pod-product-compliance
Lightning Source LLC
Chambersburg PA
CBHW080648180526
45168CB00008B/3344